Intel Galileo Blueprints

Discover the true potential of the Intel Galileo board for building exciting projects in various domains such as home automation and robotics

Marco Schwartz

PUBLISHING

BIRMINGHAM - MUMBAI

Intel Galileo Blueprints

Copyright © 2015 Packt Publishing

First published: June 2015

Production reference: 1230615

Published by Packt Publishing Ltd.
Livery Place
35 Livery Street
Birmingham B3 2PB, UK.

ISBN 978-1-78528-142-6

www.packtpub.com

Credits

Author
Marco Schwartz

Reviewers
Adam Pasztory

Alan Plotko

Christoph Schultz

Commissioning Editor
Edward Bowkett

Acquisition Editor
Harsha Bharwani

Content Development Editor
Anish Sukumaran

Technical Editor
Vivek Pala

Copy Editor
Pranjali Chury

Project Coordinator
Mary Alex

Proofreader
Safis Editing

Indexer
Monica Ajmera Mehta

Graphics
Disha Haria

Production Coordinator
Arvindkumar Gupta

Cover Work
Arvindkumar Gupta

About the Author

Marco Schwartz is an electrical engineer, entrepreneur, and blogger. He has a master's degree in electrical engineering and computer science from Supelec in France and a master's degree in micro engineering from EPFL in Switzerland.

Marco has more than 5 years of experience working in the domain of electrical engineering. His interests gravitate around electronics, home automation, the Arduino and Raspberry Pi platforms, open source hardware projects, and 3D printing.

He runs several websites around Arduino, including the Open Home Automation website, which is dedicated to building home automation systems using open source hardware.

He is the author of two other books, namely *Home Automation with Arduino* and *Internet of Things with the Arduino Yun*, both by Packt Publishing.

About the Reviewers

Adam Pasztory has a diverse background, including a BA in history from Duke University and a BS in computer science from San Francisco State University. Besides engineering, Adam has been involved in theater and films, and he enjoys developing software that entertains, informs, and enlightens.

He began his career at LucasArts, where he was involved in testing and localizing many classic games, including *Grim Fandango* and *Jedi Knight*. Later, he played key technical roles in several early-stage start-ups.

> I want to thank Cindy for helping me chase all my dreams.

Alan Plotko is a technology enthusiast with experience of developing across the full stack. He was first exposed to programming at the age of nine, when he discovered the "view source code" option in his browser. Coding then quickly turned into a hobby; this led him to take up computer science at university. Alan loves developing applications for the Web and always makes time for attending hackathons, which are typically weekend-long programming competitions where participants build projects from scratch to benefit the community. Alan's experience extends to Python development, various database technologies, including NoSQL, and frameworks for rapid application development. When he's not writing code, he spends his time writing stories; he is an avid writer, having previously self-published a fantasy novel.

Christoph Schultz was born in 1983 in Solingen, Germany.

Since his youth, he has been interested in making electronics projects. He started programming a simple text adventure game in BASIC on his brother's C64 when he was 8 years old. He learned programming in C, Java, and JavaScript all by himself in the following years. In these years, he also had his first contact with building electronic systems, when selecting and setting up the custom-made family PC.

His passion for programming and developing electronic systems finally lead to a diploma (Dipl.-Ing.) in electrical engineering from Ruhr University Bochum in Germany. Since then, Christoph has worked as an RF (Radio-frequency) system engineer in mobile phone transceiver development, first for Infineon Technologies and, since 2011, for the Intel Corporation.

Keeping his hobby alive, he actively participates in the growing maker movement. Though not part of the Galileo development team at Intel, he — like so many other makers working for Intel — was eager to get his hands on the Galileo development board. He, therefore, has used it along with the Intel Edison since day one for personal hobby projects.

www.PacktPub.com

Support files, eBooks, discount offers, and more

For support files and downloads related to your book, please visit www.PacktPub.com.

Did you know that Packt offers eBook versions of every book published, with PDF and ePub files available? You can upgrade to the eBook version at www.PacktPub.com and as a print book customer, you are entitled to a discount on the eBook copy. Get in touch with us at service@packtpub.com for more details.

At www.PacktPub.com, you can also read a collection of free technical articles, sign up for a range of free newsletters and receive exclusive discounts and offers on Packt books and eBooks.

https://www2.packtpub.com/books/subscription/packtlib

Do you need instant solutions to your IT questions? PacktLib is Packt's online digital book library. Here, you can search, access, and read Packt's entire library of books.

Why subscribe?

- Fully searchable across every book published by Packt
- Copy and paste, print, and bookmark content
- On demand and accessible via a web browser

Free access for Packt account holders

If you have an account with Packt at www.PacktPub.com, you can use this to access PacktLib today and view 9 entirely free books. Simply use your login credentials for immediate access.

Table of Contents

Preface

The Intel Galileo board is an amazing development board for all your DIY electronic projects. The board combines the power of an Intel processor with the simplicity of the Arduino platform. This makes it the perfect board for all sorts of projects, especially projects requiring complex interactions with cloud-based services, making it the ideal platform for Internet of Things applications.

In this book, we will start from simple projects that can be done with most Arduino boards. However, even at this point, we will use the advanced features of the Galileo board.

Later, we will use the Galileo board for more complex applications in fields such as the Internet of Things, home automation, and robotics.

What this book covers

Chapter 1, Setting Up the Galileo Board and the Development Environment, demonstrates how to completely set up the development environment to build and use all the projects that you will find in this book.

Chapter 2, Creating a Weather Measurement and Data Logging Station, covers how to use the inputs of the Intel Galileo board. As an example, we will make a simple weather measurement station that will log data on an SD card.

Chapter 3, Controlling Outputs Using the Galileo Board, covers how to control different devices that can be connected to the Galileo board, such as a servomotor.

Chapter 4, Monitoring Data Remotely, teaches you how to use the Ethernet port of the Galileo board and create a measurement station that can be accessed from your local network.

Chapter 5, Interacting with Web APIs, covers how to connect the Galileo board to the Internet and interact with Web APIs to add more functionalities to the board.

Chapter 6, Internet of Things with Intel Galileo, covers using the Galileo board to create applications in the very exciting field of Internet of Things.

Chapter 7, Controlling Your Galileo Projects from Anywhere, teaches you how to control your Galileo projects from any web browser, wherever you are in the world.

Chapter 8, Displaying the Number of Unread Gmail E-mails on an LCD Screen, lets you use what you learned so far and build an application to display the number of unread e-mails you have in your Gmail inbox on an external LCD screen.

Chapter 9, Automated Remote Gardening with Intel Galileo, covers building another application based on the Galileo board—a complete management system for garden irrigation. You will also be able to monitor it from anywhere in the world.

Chapter 10, Building a Complete Home Automation System, lets you use what you learned so far in this book to build a project in an exciting field—home automation. We will see how to use the Galileo board as the hub of a home automation system.

Chapter 11, Building a Mobile Robot Controlled by the Intel Galileo Board, demonstrates the use of the Galileo board as the "brain" of a mobile robot.

Chapter 12, Controlling the Galileo Board from the Web in Real Time Using MQTT, will let you discover the MQTT protocol that we will use to control the board in real time from a web browser.

What you need for this book

For this entire book, you will need an Intel Galileo board. In the first chapter of this book, you will learn how to install all the required software to configure your board.

You will also need a computer running Windows, OS X, or Linux, as this will be needed to configure your Galileo board.

Who this book is for

This book is intended for those who want to build exciting projects using the Intel Galileo board. For example, it is for people who are already experienced in using more classic Arduino boards, and want to extend their knowledge to the Intel Galileo board.

It is also for people who want to learn about electronics and programming; the Intel Galileo is the perfect platform for this.

Conventions

In this book, you will find a number of text styles that distinguish between different kinds of information. Here are some examples of these styles and an explanation of their meaning.

Code words in text, database table names, folder names, filenames, file extensions, pathnames, dummy URLs, user input, and Twitter handles are shown as follows: "To install this Arduino library, simply place the whole library folder inside the /libraries folder of your main Arduino installation folder."

A block of code is set as follows:

```
var pubnub = require("pubnub")({
    ssl              : true,  // <- enable TLS Tunneling over TCP
    publish_key      : "your_publish_key",
    subscribe_key    : "your_subscribe_key"
});
```

Any command-line input or output is written as follows:

```
unzip ngrok.zip
```

New terms and **important words** are shown in bold. Words that you see on the screen, for example, in menus or dialog boxes, appear in the text like this: "We will configure our data source here. Select **Dweet.io** and type in the name of your device."

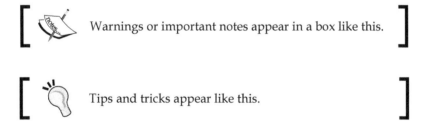

Warnings or important notes appear in a box like this.

Tips and tricks appear like this.

Reader feedback

Feedback from our readers is always welcome. Let us know what you think about this book—what you liked or disliked. Reader feedback is important for us as it helps us develop titles that you will really get the most out of.

To send us general feedback, simply e-mail feedback@packtpub.com, and mention the book's title in the subject of your message.

If there is a topic that you have expertise in and you are interested in either writing or contributing to a book, see our author guide at www.packtpub.com/authors.

Customer support

Now that you are the proud owner of a Packt book, we have a number of things to help you to get the most from your purchase.

Downloading the example code

You can download the example code files from your account at http://www.packtpub.com for all the Packt Publishing books you have purchased. If you purchased this book elsewhere, you can visit http://www.packtpub.com/support and register to have the files e-mailed directly to you.

Errata

Although we have taken every care to ensure the accuracy of our content, mistakes do happen. If you find a mistake in one of our books—maybe a mistake in the text or the code—we would be grateful if you could report this to us. By doing so, you can save other readers from frustration and help us improve subsequent versions of this book. If you find any errata, please report them by visiting http://www.packtpub.com/submit-errata, selecting your book, clicking on the **Errata Submission Form** link, and entering the details of your errata. Once your errata are verified, your submission will be accepted and the errata will be uploaded to our website or added to any list of existing errata under the Errata section of that title.

To view the previously submitted errata, go to https://www.packtpub.com/books/content/support and enter the name of the book in the search field. The required information will appear under the **Errata** section.

Piracy

Piracy of copyrighted material on the Internet is an ongoing problem across all media. At Packt, we take the protection of our copyright and licenses very seriously. If you come across any illegal copies of our works in any form on the Internet, please provide us with the location address or website name immediately so that we can pursue a remedy.

Please contact us at copyright@packtpub.com with a link to the suspected pirated material.

We appreciate your help in protecting our authors and our ability to bring you valuable content.

Questions

If you have a problem with any aspect of this book, you can contact us at questions@packtpub.com, and we will do our best to address the problem.

1
Setting Up the Galileo Board and the Development Environment

Intel Galileo Blueprints is for Arduino and electronics hobbyists who want to bring their electronic **Do It Yourself** (**DIY**) projects to the next level, using an Intel-based Arduino board — the Intel Galileo.

This book will teach you how to develop the Galileo software and how to connect the sensors for the board. It will be your guide on how to integrate the board into an **Internet of Things framework**. Indeed, many of the projects you will find in this book will be about how to connect your Galileo board to web services and monitor it remotely.

It will teach you how to create applications involving mobile robot control, home automation, remote data monitoring, and much more. This book will help you in the first steps of your Galileo projects and it will lead you closer to your mission of making great electronic creations for the world.

In this chapter, you will learn:

- Introduction to Arduino
- The Intel Galileo board
- Setting up the development environment

What is Arduino?

Arduino is an open-source single-board microcontroller, which is used in building electronics projects. It can be connected to sensors, LEDs, motors, and other devices to create an interactive display, analysis kits, and anything else electronics-based that you can think of.

Arduino has been popular among students, hobbyists, and even professionals ever since it was first introduced in 2005.

The first Arduino started as a project for the students of the Interaction Design Institute Ivrea. A hardware thesis was contributed for Hernando Barragan's (a Columbian student) wiring design. When the wiring platform was completed, a team of researchers and developers worked on the thesis to create a lighter and less expensive prototype to be available to the open source community. The five-man team that created the prototype Arduino board was led by Massimo Banzi.

Types of Arduino boards

Before the Intel-based boards, the Arduino platform was composed of either an 8-bit Atmel AVR microcontroller or a 32-bit Atmel ARM. The latest Arduino models have 6 analog inputs and 14 digital I/O pins with a USB interface.

However, as with all developing technologies, Arduino boards have evolved a lot. Here are the Arduino versions that are commercially available along with their basic features:

Name	Release date	Processor	I/O		
		Processor	Digital I/O (pins)	Digital I/O with PWM (pins)	Analog input (pins)
Arduino Zero	May 15, 2014	ATSAMD21G18A(Cortex-M0+)	14	6	6
Intel Galileo	October 17, 2015	Intel® Quark SoC X1000 Application Processor	14	6	6
Arduino Yún	September 10, 2013	Atmega32U4 / Atheros AR9331	14	6	12
Arduino Esplora	December 10, 2012	Atmega32U4			
Arduino Micro	November 8, 2012	ATmega32U4	20	7	12

Name	Release date	Processor	I/O		
		Processor	Digital I/O (pins)	Digital I/O with PWM (pins)	Analog input (pins)
Arduino Due	Octber 22, 2012	ATSAM3X8E(Cortex-M3)	54	12	12
Arduino Leonardo	July 23, 2012	Atmega32U4	14	6	12
Arduino Mega ADK	July 13, 2011	ATmega2560[25]	54	14	16
Arduino Ethernet	July 13, 2011	ATmega328	14	4	6
Arduino Uno	September 24, 2010	ATmega328P	14	6	6
Arduino Mega2560	September 24, 2010	ATmega2560	54	15	16
Arduino Fio	March 18, 2010	ATmega328P	14	6	8
Arduino (Pro) Mini	August 23, 2008	ATmega168(Pro uses ATMega328)	14	6	6
LilyPad Arduino	October 17, 2007	ATmega168V or ATmega328V	14	6	6
Arduino Pro		ATmega168 or ATmega328	14	6	6

Types of Arduino Boards

Each of these board versions differ in voltage, processor frequency, dimensions, flash memory, Electrically Erasable Programmable Read-Only Memory (EEPROM), and Static Random Access Memory (SRAM).

Among these Arduino versions, the Galileo Board is the first board to be based on Intel x86 architecture.

The Intel Galileo board

Galileo is based on 32-bit Quark SoC x1000, Intel's first ever product from the Intel Quark family, which features low power small-core products. The board is hardware compatible with Arduino's expansion cards (called shields) and is software compatible with Arduino's **Integrated software Development Environment** (IDE). It can run on Linux, Mac OSX, and Microsoft Windows.

Currently, there are two versions of Intel Galileo—Gen 1 and Gen 2. The difference between the two lies in the **General-Purpose Input/Output** (**GPIO**) pins, the latter having improved performance and compatibility with Arduino shields and accessories.

Here is a picture of the Intel Galileo Gen 2 board:

You will notice that Intel Galileo has 3.3V and 5V power supply ports, 14 digital pins, 6 analog pins, In-Circuit Serial Programming (ICSP) header, and Universal Asynchronous Receiver/Transmitter (UART) ports. Galileo has similar basic port locations as the Arduino Uno R3, but it features components that stretch its capabilities beyond the usual Arduino shield system. You can find a microSD slot, RS-232 serial port, USB port, a full-size mini-PC Express slot, Joint Test Action Group (JTAG) header, an 8-byte NOR flash, and a 100Mb Ethernet port.

Here is a short description of the Intel Galileo facility features:

- The default operating voltage of Galileo is 3.3V, but you can modify the jumper on the board to translate the voltage to 5V. The 5V option allows you to be able to use 5V Uno shields. Ensure that you check the state of this jumper before starting any project using the board.

- The 400MHz 32-bit Intel microprocessor is simple to program, can support **Advanced Configuration and Power Interface** (**ACPI**) compatible CPU sleep states and runs on an integrated **Real-Time Clock** (**RTC**).

- Among the 14 digital I/O ports, you can utilize 6 as **Pulse Width Modulation (PWM)** outputs. PWM is used to control the power sent to electrical devices: for example, to modulate the intensity of a LED or the speed of a DC motor. The analog pins have 12-bit resolution via an AD7298 A-to-D converter.

- The **6-pin ICSP header** is situated on the same location as the other Arduino boards, so you can easily plug existing shields. The ICSP supports SPI communication.

- The **Serial Peripheral Interface (SPI)** port's default frequency is 4MHz to directly support the Arduino Uno shields, but you can program it to 25MHz. Note that the Galileo SPI acts as a master and cannot be a slave to other SPIs. It can be a slave to the USB client connector, though.

- The **Peripheral Component Interconnect (PCI)** Express slot is compliant with PCIe 2.0 and works with PCIe cards with an optional converter plate. Any standard module can be connected to provide Bluetooth, Wi-Fi, or cellular connection.

- The **USB 2.0** port can support up to 128 USB endpoint devices.

- The **10-pin Standard JTAG** header is present for debugging.

- **UART TTL** is available for serial communication, while a second UART is available via a 3.5mm jack for RS-232 support.

- The **USB Device** port supports serial (CDC) USB communication.

- Mice, keyboards, and other peripherals can be connected through the **USB Host** port.

- The **Ethernet RJ45** connector enables wired network connections.

- Access to the **micro SD** is through the SD library.

- Use of **TWI/I2C bus** is simplified through the Arduino wire library.

Aside from the improved shield compatibility of Galileo Gen 2, it has upgraded the input power range from 7V to 15V instead of just 5V, upgraded resolution of digital pins with PWM, and provides optional 12V PoE support.

[In this book, we will use Galileo Gen 2.]

Now that you are familiar with the parts and features of Galileo, you will learn how to set up the development environment.

Setting up the development environment

Note that you should not use the same power supply for Galileo Board Gen 1 and Gen 2. Gen 1 is only rated at 3.3V to 5V, while Gen 2 boards are powered up with a voltage between 7V and 15V. Using the Gen 2 power supply on Gen 1 boards will cause permanent damage to the Gen 1 boards.

Let's get started with setting up the development environment. You need to perform the following steps:

1. First, connect the Galileo DC jack to power up your board. You will see the LEDs light up. Then, connect your Galileo board to your computer through the USB client port. Look at the picture here for the connection reference:

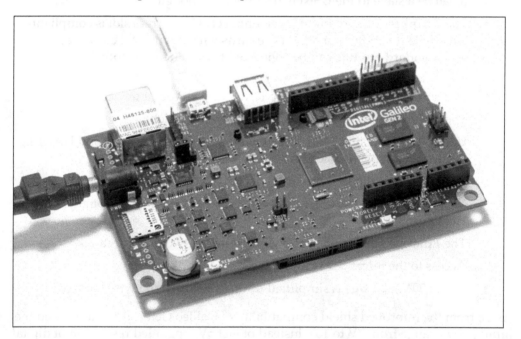

2. Next, download the Galileo development environment. It is a custom version of the usual Arduino IDE developed by Intel. You may download the IDE at the following link:

    `https://communities.intel.com/docs/DOC-22226`

3. Install the downloaded software in your computer and launch the application.

4. Before proceeding, ensure that the latest firmware is installed. To update the firmware, follow these steps:

    1. Navigate to **Help | Firmware Update**. You will need to wait for a bit, as this takes several minutes.

    2. Now, load the Blink example in the IDE by navigating to **Examples | Basics | Blink**.

5. The Blink code will load and you will end up with the editor window shown in the following screenshot:

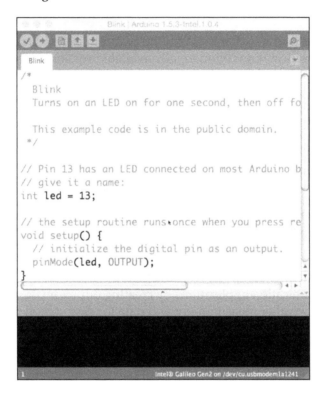

6. Click on **Upload** to load the program.
7. The LED will blink on the board.

That's how easy navigation is on the Galileo interface!

If your LED light does not blink, check the following:

1. Confirm that you have the latest Galileo IDE loaded.
2. Ensure that the latest firmware is installed.
3. Check the version of the board that has been selected by navigating to the board in **Tools** | **Board**.
4. Ensure that there are no loose cable connections.

Summary

In this chapter, you learned what the Arduino platform is, how it came to be, what it is for, where it is used, and its commercially available versions. Most of all, we found out about one of the latest boards of the Arduino platform and the first to use the Intel microprocessor: the Intel Galileo board.

We know that there are two current versions of Intel Galileo: Gen 1 and Gen 2. We will focus on Galileo Gen 2 in this book, but Gen 1 will work as well for all the projects of the book. You also learned how to set up the development environment for Intel Galileo and how to update its firmware.

We are just starting to explore the first ever Arduino board to run on an Intel microprocessor and we are just barely scratching the surface. You have so much more to discover and learn!

In the following chapters, you will learn more of Galileo's basic and core functions. You will study structure, variables, and functions of Gen 2. We will explore digital, analog, and communication applications by creating our own DIY projects. We'll also start with the weather measurement and data logging station project in the next chapter.

2

Creating a Weather Measurement and Data Logging Station

In this chapter, you will learn how to create a weather measurement and data logging station. Through this project, we will explore how to use the inputs of the Intel Galileo board to measure data from sensors. We will connect analog and digital sensors to the board. The sensors will measure the temperature, humidity, and ambient light. We will then log these measurements on to the SD card.

Let's dive in!

Hardware and software requirements

Firstly, you need to secure your Intel Galileo board. This will serve as our microcontroller board. Note that we'll use Gen 2 for this project.

We'll only use two sensors — a DHT11 and a photocell. We will need the DHT11 for the temperature and humidity measurements and the photocell for measuring the ambient light level.

Aside from these major components, you will also need a 1N4148 diode, 4.7K and 10K Ohm resistors, breadboard, jumper wires, and, of course, an SD card.

Here is a list of all the components needed for this project:

- Intel Galileo board (`https://www.adafruit.com/products/2188`)
- DHT11 sensor + 4.7K Ohm resistor (`http://www.adafruit.com/product/386`)

- 1N4148 diode (https://www.sparkfun.com/products/8588)
- Photocell (https://www.adafruit.com/products/161)
- 10K Ohm resistor (https://www.sparkfun.com/products/8374)
- Breadboard (https://www.adafruit.com/products/64)
- Jumper wires (https://www.adafruit.com/products/1957)
- MicroSD card

At this point, we're ready to start setting up the Galileo board and the development environment. If you need to review how this is to be done, you can go back to *Chapter 1, Setting Up the Galileo Board and the Development Environment.*

Also, ensure that you have installed the following DHT files in your Arduino library:

https://github.com/marcoschwartz/DHT_Galileo

To install a library, you can follow this guide:

http://arduino.cc/en/Guide/Libraries

Configuring the hardware

After all the required files are installed, we can start configuring the hardware by following these steps:

1. Start with the DHT11 sensor, place it on the breadboard.
2. Connect pin 1 to VCC (5V) on the Galileo board and pin 4 to GND on the Galileo board.
3. Connect the sensor's pin 2 to both Galileo pin 5 and the diode.
4. Connect the other end of the diode, and the cathode, to the Galileo board pin 6.
5. Place one leg of the photocell in series with the 10k Ohm resistor.
6. The other end of the resistor should be grounded by connecting it to the GND pin or the blue power rail.
7. Connect the row where the resistor and the photocell meet to the A0 pin of the Galileo board.
8. Connect the other leg of the photocell to VCC.
9. Insert the SD card into Galileo's onboard SD card reader on the side of the board.

You can refer to the following schematic as a guide:

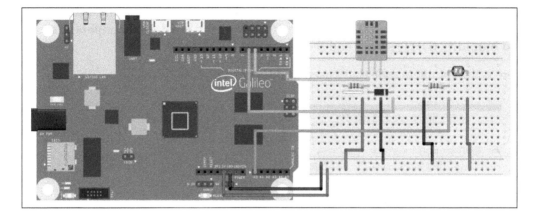

Here is a picture of the hardware that was configured:

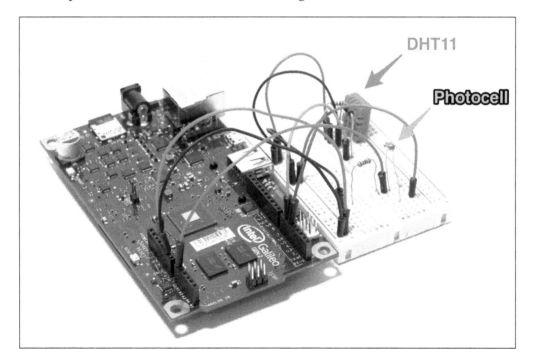

Testing the sensors

After setting up the hardware, we can now test whether the sensors are working properly. To do so, we need to write a code to test both the **digital temperature and humidity (DHT)** sensor and the photocell.

Here is the complete code that I used in this project:

```
// Libraries
#include "DHT_Galileo.h"

// DHT sensor type
#define DHTTYPE DHT11 // DHT 11

// DHT sensor pins
#define DHTIN 5
#define DHTOUT 6

// DHT instance
DHT_Galileodht(DHTIN,DHTOUT, DHTTYPE);

void setup()
{
  // Initialize the Serial port
Serial.begin(115200);

  // Init DHT
dht.begin();
}

void loop()
{
  // Measure from DHT
float temperature = dht.readTemperature();
float humidity = dht.readHumidity();

  // Measure light level
floatsensor_reading = analogRead(A0);
float light = sensor_reading/1024*100;

  // Display temperature
Serial.print("Temperature: ");
Serial.print((int)temperature);
Serial.println(" C");
```

```
    // Display humidity
Serial.print("Humidity: ");
Serial.print(humidity);
Serial.println("%");

    // Display light level
Serial.print("Light: ");
Serial.print(light);
Serial.println("%");
Serial.println("");

    // Wait 500 ms
delay(500);
}
```

Downloading the example code

You can download the example code files from your account at
http://www.packtpub.com for all the Packt Publishing books
you have purchased. If you purchased this book elsewhere, you
can visit http://www.packtpub.com/support and register
to have the files e-mailed directly to you.

Following are the steps that we perform with the execution of the previous code:

1. We first need to include the DHT sensor library:

   ```
   #include "DHT_Galileo.h"
   ```

2. Then, we define the sensor type:

   ```
   #define DHTTYPE DHT11 // DHT 11
   ```

3. Define the pins of the DHT sensor:

   ```
   #define DHTIN 5
   #define DHTOUT 6
   ```

4. After this, we create the DHT object:

   ```
   DHT_Galileodht(DHTIN,DHTOUT, DHTTYPE);
   ```

5. In the setup() function, we initialize the Serial port:

   ```
   Serial.begin(115200);
   ```

6. We also initialize the sensor:

   ```
   dht.begin();.
   ```

7. In the `loop()` function, we take measurements from the DHT sensor:

    ```
    float temperature = dht.readTemperature();
    float humidity = dht.readHumidity();
    ```

8. Also, we also take light measurements:

    ```
    floatsensor_reading = analogRead(A0);
    float light = sensor_reading/1024*100;
    ```

9. Then, we print the temperature readings:

    ```
    Serial.print("Temperature: ");
    Serial.print((int)temperature);
    Serial.println(" C");
    ```

10. We also print the humidity readings:

    ```
    Serial.print("Humidity: ");
    Serial.print(humidity);
    Serial.println("%");
    ```

11. We next print the light levels:

    ```
    Serial.print("Light: ");
    Serial.print(light);
    Serial.println("%");
    Serial.println("");
    ```

12. And we wait for 500 ms between measurements:

    ```
    delay(500);
    ```

 This code can be downloaded from the following link:
https://github.com/marcoschwartz/galileo-blueprints-book

Upload the code into the Galileo board (refer to *Chapter 1, Setting Up the Galileo Board and the Development Environment*). After uploading, open the serial monitor to check out the resulting display.

You will see the following results on your display:

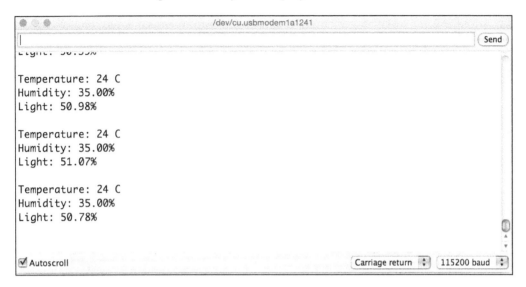

Logging data onto the SD card

Now that data from the sensors is being displayed on the serial monitor, we will log the results on an SD card. You need the following code to make this work.

You will notice that the following code is similar to the previous code; only a few parts are added. This is the complete code for this recipe:

```
// Libraries
#include "DHT_Galileo.h"
#include <SD.h>

// DHT sensor type
#define DHTTYPE DHT11 // DHT 11

// DHT sensor pins
#define DHTIN 5
#define DHTOUT 6

// DHT instance
DHT_Galileodht(DHTIN,DHTOUT, DHTTYPE);

void setup()
{
  // Initialize the Serial port
```

```
Serial.begin(115200);

  // Init SD card
Serial.print("Initializing SD card...");

if (!SD.begin()) {
Serial.println("Card failed, or not present");
return;
  }
Serial.println("card initialized.");
system("/sbin/fdisk -l > /dev/ttyGS0");

  // Init DHT
dht.begin();

  // Set date and time (only run once!)
  //system("date 171110202014"); //sets the date & time to 10:00 1st
Jan 2014
}
void loop()
{
  // Measure from DHT
int temperature = dht.readTemperature();
int humidity = dht.readHumidity();

  // Measure light level
floatsensor_reading = analogRead(A0);
int light = sensor_reading/1024*100;

  // Prepare data to be logged
  String dataString = "";

  // Get time
system("date '+%H:%M:%S' > /home/root/time.txt");
  FILE *fp;
fp = fopen("/home/root/time.txt", "r");
charbuf[9];
fgets(buf, 9, fp);
fclose(fp);
dataString += String(buf) + ",";

  // Add measurements
dataString += String(temperature) + ",";
dataString += String(humidity) + ",";
```

```
dataString += String(light);

    // Log data
    File dataObject = SD.open("datalog.txt", FILE_WRITE);

    // Check temperature range to avoid sensor errors
if (temperature > -80 && temperature < 80){

      // If the file is available, write to it:
if (dataFile) {
dataFile.println(dataString);
dataFile.close();

        // Print to the serial port too:
Serial.print("Logged data: ");
Serial.println(dataString);
      }

      // if the file isn't open, pop up an error:
else {
Serial.println("Error opening datalog.txt");
      }
    }

    // Wait 10 seconds between measurements
delay(10000);
}
```

Let's now see the details of this code. First, we import the following libraries:

```
#include "DHT_Galileo.h"
#include <SD.h>
```

Then, we initialize the SD card:

```
Serial.print("Initializing SD card...");

if (!SD.begin()) {
Serial.println("Card failed, or not present");
return;
    }
Serial.println("card initialized.");
system("/sbin/fdisk -l > /dev/ttyGS0");
```

Next, we set the date. This needs only needs to be done once because the board will remember the date and count from this date using the internal clock. After setting it once, it should be commented out:

```
system("date 171110202014");
```

We measure data in the `loop()` function:

```
// Measure from DHT
int temperature = dht.readTemperature();
int humidity = dht.readHumidity();

  // Measure light level
floatsensor_reading = analogRead(A0);
int light = sensor_reading/1024*100;
```

Also, we create a string that will contain data:

```
String dataString = "";
```

Now, we get the time from the Galileo board and write it in a file called time.txt:

```
system("date '+%H:%M:%S' > /home/root/time.txt");
FILE *fp;
fp = fopen("/home/root/time.txt", "r");
charbuf[9];
fgets(buf, 9, fp);
fclose(fp);
dataString += String(buf) + ",";
```

We also add the measurements:

```
dataString += String(temperature) + ",";
dataString += String(humidity) + ",";
dataString += String(light);
```

This creates a new file on the SD card:

```
File dataFile = SD.open("datalog.txt", FILE_WRITE);
```

Then, we check whether the temperature is valid. If it is, the data will be logged:

```
if (temperature > -80 && temperature < 80){

    // If the file is available, write to it:
if (dataFile) {
dataFile.println(dataString);
```

```
dataFile.close();

    // Print to the serial port too:
Serial.print("Logged data: ");
Serial.println(dataString);
    }

    // if the file isn't open, pop up an error:
else {
Serial.println("Error opening datalog.txt");
    }
  }
```

We do this every 10 seconds:

```
delay(10000);
```

Once you have uploaded the code, you can open the serial monitor to check the results. You will get a display like this:

Since we have stored the data on the SD card, we can remove the card and transfer the text data on our computer. (Look for `datalog.txt` in the root folder.) It can even be imported to Excel:

	A	B	C	D	E	F	G	H
1	10:20:07	24	35	61				
2	10:20:27	24	35	61				
3	10:20:48	24	35	61				
4	10:20:58	24	35	61				
5	10:21:09	24	35	61				
6	10:21:19	24	35	62				
7	10:21:29	24	35	62				
8	10:21:40	24	35	62				

Once data is imported and tabulated in Excel, we can plot the data using the Excel graph functionalities that you can find in the main toolbar. Here is a sample of the result plotting:

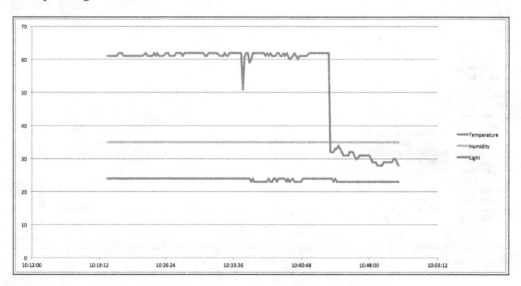

Great! You can now build your own weather and data logging system.

Summary

In this chapter, we first built the hardware by connecting the sensors and SD card to the Galileo board. We installed the necessary files and code on the Galileo to make it work as desired. Then, we tested the sensors and displayed the measurements on the serial monitor. We read the data and stored it on the SD card. Lastly, we analyzed and plotted the data in Excel.

To go further with this project, you can add more sensors, such as precipitation and wind speed detectors.

In the next chapter, you will learn how to control the outputs of the Galileo board. We will connect actuators, such as relays and servomotors, to the board and control them via push buttons and potentiometers.

Controlling Outputs Using the Galileo Board

3

By now we have already familiarized ourselves with the Galileo board. You learned how to set up its development environment and explore configuring its analog inputs, and how to use digital sensors. We will now move on to exploiting the output capabilities of the board.

In this chapter, you will learn how to configure the output of the Galileo board. Specifically, we will play with its precise 12-bit PWM output to integrate and control actuators, such as sensors and servomotors through it. On the software side, we will write code that will control these actuators through push buttons and potentiometers.

This chapter is divided into two parts. In the first part, we will discuss how to control the relay through a push button. In the second part, we will take a look at how to control the servomotor with a potentiometer.

Let's dive in!

Hardware and software requirements

Before we proceed, ensure that you have everything needed to build the two mini projects.

First, we will need the Galileo board. Again, all the projects in this book utilize the Galileo board Gen 2.

We will use two actuators: a **relay module** and a **servomotor**. The relay module used in this project is a Pololu 5V **single-pole, double-throw (SPDT)** relay. The module already includes a basic carrier PCB, terminal blocks for the switch connectors, and a straight male header for the controls.

Meanwhile, the servomotor used is a simple continuous full-rotation servo motor. This servo is rated at 4.8V – 6V. However, you can use any servomotor for this project.

We will also need a potentiometer and a push button to control the actuators. The **potentiometer** used in this project is a center-tap linear 10k Ohm type. The push button is a **single-pole single-throw** (**SPST**) switch rated at 50mA.

The other things that we will need to complete our project are: resistors, a breadboard, and male/male jumper wires.

For your convenience, you can refer to the list of all components used in this chapter here:

- Intel Galileo Gen 2 board (https://www.sparkfun.com/products/13096)
- Pololu 5V relay module (http://www.pololu.com/product/2480)
- Servomotor (https://www.sparkfun.com/products/9347)
- 10k Ohm potentiometer (https://www.sparkfun.com/products/9288)
- Push button (https://www.sparkfun.com/products/97)
- 1k-10k Ohm resistor (https://www.sparkfun.com/products/8374)
- Breadboard (https://www.sparkfun.com/products/12002)
- Male/male and male/female jumper wires (https://www.sparkfun.com/products/9387)

Assembling the relay controller

Let's start with the first part of this chapter—the relay controller. We are using a relay here to be able to control devices using high currents. Indeed, from the Galileo board alone we can only control devices that use low currents, such as LEDs.

We will assemble it (the relay controller) first. The circuit is designed in such a way that the push button will control the relay. Here is a schematic for your reference:

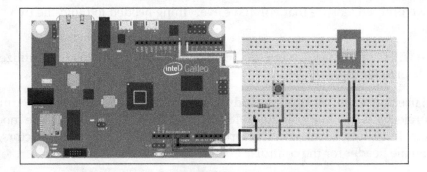

This circuit is easy to build. Following are the steps which will guide you to do this:

1. First, connect the VCC and GND to the top horizontal rows of the breadboard.
2. Then, connect one side of the push button to the VCC.
3. The other side of the switch should be connected to one leg of the resistor.
4. The other leg of the resistor should be grounded.
5. Connect the other side of the switch to pin 8 of the Galileo board (you can refer to the switch reference in the list of components if you want to view its internal connections).
6. As for the relay, connect the VCC pin to one of the breadboard's top horizontal rows, where the VCC of the Galileo is connected.
7. Then, connect the GND pin of the relay to the GND row of the breadboard.
8. The EN pin of the relay should go to pin 7 of the Galileo board.

And you're done!

After connecting everything, this is how the relay controller circuit connection looked on my breadboard:

Using the relay controller

Now that we have assembled the relay controller, we will build the Arduino sketch to make the circuit do as we like. We will create the sketch in such a way that it will be able to control the relay through the push button. To do so, you will need to follow the code given in these steps:

1. First, we will declare the relay and the button pins. The relay is pin 7, while the button is pin 8. To do this, follow the code:

    ```
    int relayPin = 7;
    int buttonPin = 8;
    ```

2. Next, we will create variables for the button states:

    ```
    int buttonState;
    int lastButtonState = LOW;
    int previousButtonState = LOW;
    ```

3. To debounce the button and achieve a stable state for the relay, we will add the following code. We will declare a default last debounce time and debounce delay:

    ```
    long lastDebounceTime = 0;
    long debounceDelay = 50;
    ```

4. Then, we will create a variable for the state of the relay:

    ```
    int relayState = LOW;
    ```

    Inside the `setup()` function of the sketch, we will set the pins for the relay and the button. The button switch will be an input, while the relay pin will be an output. Here is the code for this:

    ```
    pinMode(relayPin, OUTPUT);
    pinMode(buttonPin, INPUT);
    ```

    In the `loop()` function, we are going to read data from the sensor, which is the button pin:

    ```
    int reading = digitalRead(buttonPin);
    ```

5. If the pushbutton is pressed, we will debounce the button and switch to the relay state. To do this, we will create a debouncing timer that will figure out whether the last debounce time is greater than the debounce delay.

    If it is, it will check whether the button state has changed. If it has, we need to switch to the relay state. Here is the code for this:

```
if (reading != lastButtonState) {

// Reset the debouncing timer
lastDebounceTime = millis();
}

// Debounce code (need to get a clear reading of the button state)
if ((millis() - lastDebounceTime) > debounceDelay) {

// If the button state has changed
if (reading != buttonState) {
buttonState = reading;
}

// If the button was pressed, changed the output
if (previousButtonState == LOW &&  buttonState == HIGH) {
relayState = !relayState;
}
}
```

6. Next, by adding the following code, we will apply the right state to the pin that controls the relay:

```
digitalWrite(relayPin, relayState);
```

7. We will also update the previous states of the button using the following code:

```
previousButtonState = buttonState;
lastButtonState = reading;
```

This is all the code we need for the relay controller.

 Note that you can find the whole code for this section inside the GitHub repository of this book:

```
https://github.com/marcoschwartz/galileo-
blueprints-book
```

8. After you have downloaded the code, just upload it to your Galileo board.

9. After uploading the code, you can test the code to check whether it works alongside the relay controller. To do so, just press the pushbutton. Upon pressing, the relay should switch on. Press the button again to switch the state of the relay. This time, the switch should turn off.

If it did, congratulations! You have completed the first half of this project.

However, if the circuit didn't work as desired, you can do the following checks to debug your work:

1. First, go back to the breadboard connection. Ensure that everything is connected correctly. You can go back to the previous schematic and instructions to check your circuit.

2. Next, you might also want to check whether the code uploaded to your Galileo board is the one that you downloaded from the GitHub repository of this book.

Assembling the servomotor controller

We now proceed to the second part of this project—the **servomotor** controller. By this time, we have an idea on how to use the output pins of the Galileo board.

We will first assemble the servomotor controller circuit for this project.

Here is the schematic that you can refer to as your guide:

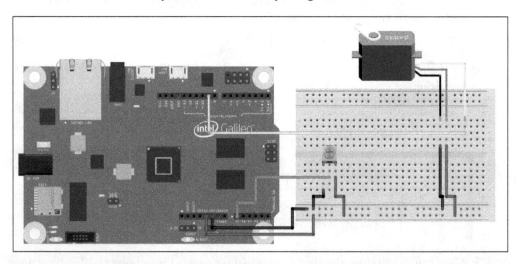

Aside from the schematic representation, here is a step-by-step guide to make the servomotor controller. The steps are carefully laid out for you to be able to finish the circuit in the easiest way possible:

1. First, you have to place the potentiometer on the far end of the breadboard. This will be our starting-point.

2. Then, connect one of the potentiometer's pins to VCC, another pin to GND, and its middle pin to the analog pin A0 of the Galileo board (you can check out the link of the potentiometer in the hardware requirements section portion, to see the internal connections through its datasheet).That's all the connections we need for the potentiometer.

3. Next, place the servomotor on the opposite end of the breadboard.

4. Get the black wire of the servomotor and connect it to the GND horizontal row of the breadboard.

5. Now, get the red wire and connect it to one of the VCC horizontal row pins of the breadboard.

6. Then, connect the servomotor's last wire, which is colored blue green to pin 9 of the Galileo board.

That's every connection that we need for the servomotor controller.

Here is how my servomotor controller circuit looked after all the connections were made:

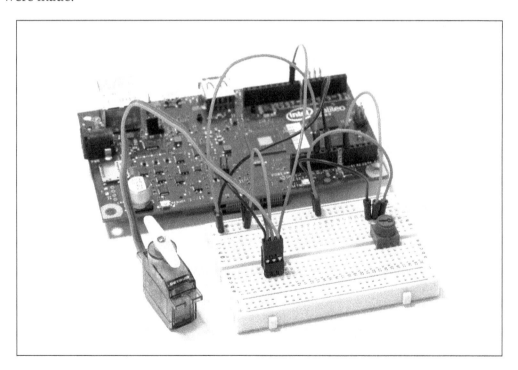

Using the servomotor controller

After we have assembled the servomotor controller circuit, we need to work on the Arduino sketch of the servomotor controller.

However, before that, you need to know that we need to match the turn of our potentiometer to the change on the angle of the servomotor. We should be wary of the potentiometer and be extra careful of our precision. We want our servomotor to give a relevant change with the turn that we make on our potentiometer.

Since we are using the Galileo board Gen 2, this wouldn't be a problem. Thanks to its 12-bit PWM output, we can be ensured of precision control.

Galileo Gen 2 makes use of an NXP PCA9685 PWM driver IC with 12-bit resolution. This is what makes a fine-grained control on the PWM duty cycle possible.

At this time, we are ready to write the code which will enable us to control the servomotor through the potentiometer, by following these steps:

1. The first thing that we should do is include the servo library, so that we will have codes that will be readily available to control our servomotor:

    ```
    #include <Servo.h>
    ```

2. Then, we will declare A0 as the potentiometer pin:

    ```
    int potPin = A0;
    ```

3. Next, create the servo object. You can name it whatever you like. In this project, `myservo` is the name of the object:

    ```
    Servo myservo;
    ```

4. In `setup()`, we will set the potentiometer pin as our input:

    ```
    pinMode(potPin, INPUT);
    ```

5. Now attach the servo to pin 9:

    ```
    myservo.attach(9);
    ```

6. In the `loop()` function, we will read data from the analog pin A0. After this, we will convert the analog result in to an angle so that the servo will understand it. To do so, we will divide the reading by *1023* and multiply it by *180*, to obtain the result in degrees:

    ```
    float reading = analogRead(potPin);
    float servo_position = reading/1023. * 180.;
    ```

7. Finally, we apply this angle to the servomotor by writing this line:

    ```
    myservo.write(servo_position);
    ```

This sums up all the code that we need for the servomotor controller.

 You can find the entire code for this section inside the GitHub repository of this book. You can go to the following link to download a copy:

`https://github.com/marcoschwartz/galileo-blueprints-book`

Once you have downloaded it, you can upload the sketch to your Galileo board.

To test whether everything is working, you can try twisting the head of the potentiometer. When you do so, you should be able to see the horn of the servomotor move. The servo should follow instantly since we are using the Galileo board Gen2, which, as previously pointed out in this chapter, is very precise.

If the servomotor's horn fails to move when you have already turned the head of the potentiometer, you can check on a couple of things to debug your project:

1. First, go back to the servomotor schematic and check all your connections. You may have just missed one thing.

2. If your circuit is perfectly connected, you should check whether the code that you have uploaded to your Galileo board is the correct one.

Summary

In this chapter we discussed two mini projects, which involved configuring the outputs of the Galileo Board 2. We built a relay controller and a servomotor controller. We also created Arduino sketches for each project and uploaded them into the Galileo board. In the relay controller, we installed a push button, which controlled the relay switch. For the servomotor controller, we installed a linear potentiometer that controlled our servomotor.

To go further in exploring how the output of the Galileo board 2 works, you can make use of other output. For example, you can use a couple of light-emitting diodes (LEDs) or DC motors.

In the next chapter, you will learn how to monitor data remotely using the Ethernet connection of the Galileo Gen 2. It will teach you how to create a measurement station by using an Ethernet port, and how to access it via a network.

4
Monitoring Data Remotely

In the previous chapter, we only used the basic functions of the board via the Arduino IDE, and specifically learned how to configure the outputs of the Galileo. You will now learn how to monitor data remotely using the board.

In this chapter, you will learn how to use the onboard Linux machine to get the full power of the board (yes, the board can be configured such that its hardware can be manipulated using the Linux operating system).

This project will also be the first one in which we will explore the onboard Ethernet port. The port connects the board up to any 10/100 Mbps LAN.

All these may sound complicated, but the steps are easy to follow. We will first configure the Linux machine. Then, we will install the Intel XDK software to configure the board remotely.

Finally, we will put everything in action with a simple data monitoring project. Let's start!

Hardware and software requirements

First, you need to get all the required components for this project.

Here are the details of the different components used in this chapter's project:

- The Ethernet connection for the Galileo board is made through the RJ45 connector, so ensure that you have a RJ45 cable around.

- We will also use a TMP36 sensor that is a low voltage (2.7V to 5.5V) precision Centigrade temperature sensor. It has a scale factor of 10mV/ degree Celsius and an accuracy of +/- 2 degree Celsius over temperature.

- The photocell used in this project has a light resistance of about 1k Ohm and a dark resistance of about 10k Ohm.

- The other components that we will need are breadboard, male/male jumper wires, and a MicroSD card that is at least 4GB.

Here is a list of all the components used in this chapter:

- Intel Galileo Gen 2 board (`https://www.sparkfun.com/products/13096`)
- TMP36 sensor (`https://www.sparkfun.com/products/10988`)
- Photocell (`https://www.sparkfun.com/products/9088`)
- 10K Ohm resistor (`https://www.sparkfun.com/products/8374`)
- Breadboard (`https://www.sparkfun.com/products/12002`)
- Male/male jumper wires (`https://www.sparkfun.com/products/9387`)
- MicroSD card (at least 4 GB)

On the software side of things, you will just need to download and install the Intel XDK IoT Edition.

You can find the software here:

`https://software.intel.com/en-us/html5/xdk-iot`

Installing the Linux image

The first thing that we should work on is the Linux image. We need to install the Linux image to access the board remotely. This is a special distribution to access the **Internet of Things (IoT)** functions of the board.

To do so, connect the Ethernet cable between the Galileo board and your router, as shown in the following image:

You need to perform the following steps for the installation:

1. After connecting to the Ethernet, download the Linux image from the following link:

   `https://software.intel.com/en-us/iot/downloads`

2. When the download is complete, unzip the files in a folder.

3. Then, insert the SD card in your computer via a MicroSD adapter.

4. Follow the instructions on the official Intel page, depending on your operating system:

   ○ For mac users, you can refer to this tutorial:

   `https://software.intel.com/en-us/node/530415`

   ○ For Linux users, instructions can be found here:

   `https://software.intel.com/en-us/node/532598`

   ○ Lastly, Windows users can refer to this set of instructions:

   `https://software.intel.com/en-us/node/530353`

Be patient. Installing the image can take a while. Once done, insert a MicroSD card into the board and reboot it.

We are now set with the software at this point. We will use the Intel XDK later to configure the board.

Configuring the hardware

Now that the required Linux software is in place, we can add sensors to monitor the data that we need.

Here is a schematic to help you out in configuring your own hardware:

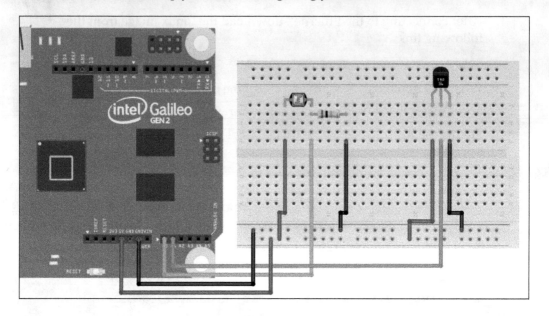

1. First, connect the 5V and GND to the top or bottom horizontal rows of the breadboard.

2. Then, connect the left pin of the TMP36 sensor to the 5V horizontal slot, the middle pin to the analog pin A1, and the right pin to the GND horizontal slot on the breadboard.

3. Then, position the photocell in such a way that it is in series with the resistor. The node where the photocell and resistor meet should be connected to the board's analog pin A0.

4. Next, connect the free leg of the photocell to the 5V line. The free leg of the resistor should be pinned to the GND.

Here is how the hardware configuration looks when everything is connected:

That's it. We are now ready to use the project.

Accessing measurements remotely

What we want to do now is access the measurements via the Ethernet port. We will not use the Arduino IDE here. We will use the Intel XDK IDE, and program the board using Node.js. Node.js makes creating and configuring apps running on these IoT devices very easy. Node.js also enables you to connect IoT sensors through JavaScript programming. You can find more information about Node.js here:

`http://nodejs.org/`

You can also check out *Mastering Node.js* by Packt Publishing.

We can now start programming the board with JavaScript.

The Intel XDK IDE lets you create, debug, and deploy Node.js applications on your IoT device. In this case, it's our Galileo board.

This is what you will see on starting the XDK:

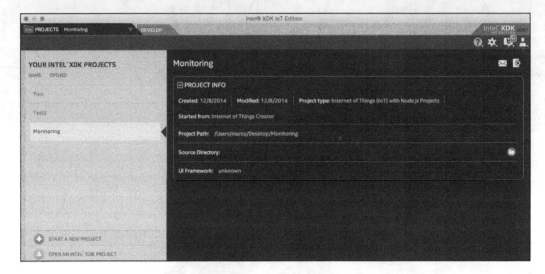

First, create a new project. You can use one of the example projects to create all the files that you require, for example, the 'Local Temperature' project.

Then go to the **DEVELOP** panel. You will see the following main window:

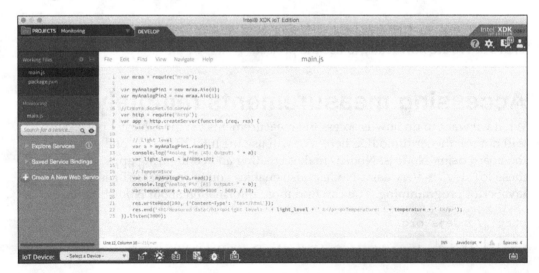

Let's now take a look at the complete code used in this part:

```
var mraa = require("mraa");

var light_sensor_pin = new mraa.Aio(0);
var temp_sensor_pin = new mraa.Aio(1);

//Create server
var http = require('http');
var app = http.createServer(function (req, res) {
    'use strict';

    // Light level
    var a = light_sensor_pin.read();
    console.log("Analog Pin (A0) Output: " + a);
    var light_level = a/4096*100;

    // Temperature
    var b = temp_sensor_pin.read();
    console.log("Analog Pin (A1) Output: " + b);
    var temperature = (b/4096*5000 - 500) / 10;

    res.writeHead(200, {'Content-Type': 'text/html'});
    res.end('<h1>Measured data</h1><p>Light level: ' + light_level + '
%</p><p>Temperature: ' + temperature + ' C</p>');
}).listen(3000);
```

Let's now break the codes down and analyze what each line does.

We need the MRAA module, which we will use to measure data from the pins, so the following line of code is used:

```
var mraa = require("mraa");
```

We will use this MRAA module a lot in this book, as it will give us access to all of the pins on the Galileo board from the Node.js server. As such, we won't have to use the Arduino IDE again.

Then, we declare one instance per pin:

```
var light_sensor_pin = new mraa.Aio(0);
var temp_sensor_pin = new mraa.Aio(1);
```

We create the server that will run on the board:

```
var http = require('http');
```

We also start our server here. Don't worry, we will close this piece of code at the end:

```
var app = http.createServer(function (req, res) {
```

After that, whenever we receive a request, we trigger the sensor to measure the light level:

```
// Light level
var a = light_sensor_pin.read();
console.log("Analog Pin (A0) Output: " + a);
var light_level = a/4096*100;
```

We calculate the temperature as well:

```
// Temperature
var b = temp_sensor_pin.read();
console.log("Analog Pin (A1) Output: " + b);
var temperature = (b/4096*5000 - 500) / 10;
```

Next, we print these measurements on a simple web page:

```
res.writeHead(200, {'Content-Type': 'text/html'});
res.end('<h1>Measured data</h1><p>Light level: ' + light_level + ' %</
p><p>Temperature: ' + temperature + ' C</p>');
```

Finally, we start the server on port 3000, which closes the statement we opened earlier:

```
}).listen(3000);
```

We use port 3000 here, just as usual web servers are using port 80. Using different ports allows you to have many apps running on a server with one given IP address.

> Note that you can download the complete code from the following link:
> https://github.com/marcoschwartz/galileo-blueprints-book

Now, we are ready to test the application.

To do so, we will add the device first. It should show up automatically in the Intel XDK:

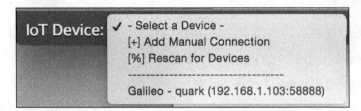

When you have found it, configure the environment as follows:

Now, click on the Build icon. Wait a bit (it takes a while), and then upload the code to the board:

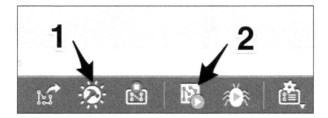

You are now ready to test the application.

Start by going to a web browser. Then, type the following:

```
ip_of_galileo_board:3000
```

You will be able to see the measurements by now. Here is an example:

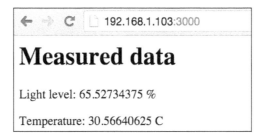

Summary

We're done! Congratulations on building your own data monitoring system which can be accessed remotely.

In this chapter, we have made use of Intel's XDK IDE and JavaScript to configure a simple remote data monitoring system. We installed a Linux image on the board, and configured the hardware to include the sensors. We also accessed the data remotely by sending commands through JavaScript. Then, we passed on the data remotely and printed this data inside a web browser.

In the next chapter, we will continue exploring IoT applications by interfacing it with online APIs, such as Twitter. Using such APIs, you will be able to broadcast your measurement results to your social media accounts.

5
Interacting with Web APIs

In the previous chapter, you learned how to use the onboard Ethernet port, communicate locally with the board, and retrieve data from it.

In this chapter, we will perform a similar experiment. The only difference is that we will interface our board with web APIs. This gives us the option to broadcast our results through social media platforms, such as Twitter.

Here is a walk-through on how we are going to go about this:

- We will structure our app using Express. **Express** is a flexible and minimalistic web app framework for Node.js, which provides a robust set of both mobile and web features. It integrates well with the Node.js platform.
- Next, we will grab weather forecasts from the Internet and integrate them in our application.
- Finally, we will use Twitter to post the data online. With this, we can regularly tweet weather updates through our Twitter account.

Let's proceed!

Using Express to structure our application

Firstly, we need to create a structure for our application. As previously mentioned, we are going to use Express to do this.

Through Express, we will separate the data acquisition from the interface. You can visit the official site given here to learn how Express works, how to use it for the first time, how to handle errors, and a lot more at http://expressjs.com/.

The measurement part will stay in Node.js. Since Node.js is an asynchronous event-driven type of framework, it can easily build fast and scalable net applications. This is what makes it efficient and lightweight to use, and a perfect choice if one wants to run real-time and data-intensive apps on multiple devices.

Meanwhile, the interface will use Jade, which is a Node.js templating engine created for HTML.

We will then link up both Node.js and Express to the Jade interface using JavaScript.

As for the hardware, we can reuse the one that we made for our data monitoring project - the one that we created in the previous chapter.

So, the software will be quite similar to the previous chapter's code, but there will be some changes that we need to make first. Let's review the code for this part.

In the `main.js` file, we need to include Express:

```
var express = require('express');
```

Then, we create the application:

```
var app = express();
```

We will now set Jade as the view engine:

```
app.set('view engine', 'jade');
app.set('views', __dirname + '/views');
```

Here, `__dirname` is the root directory of the app. In the next line, we define where the JavaScript files will be stored:

```
app.use(express.static(__dirname + '/public'));
```

Now, we will define the main route where the interface can be accessed:

```
app.get('/', function(req,res){
  res.render('interface');
});
```

Compared to earlier code, we encapsulate functions in API calls which return JSON data. JSON is a format used to exchange data that is usually found in APIs, such as the one we are creating, and to exchange data between parts of applications.

We will then simply be able to access this API by typing the desired URL in to a browser. This API will also be accessed later by the interface, using JavaScript.

For example, for temperature, we define the API access as follows:

```
app.get('/api/temperature', function(req,res){

    // Measure
    var b = temp_sensor_pin.read();
    console.log("Analog Pin (A1) Output: " + b);
    var temperature = (b/4096*5000 - 500) / 10;

    // Send answer
    json_answer = {};
    json_answer.temperature = temperature;
    res.json(json_answer);
});
```

We then start the app at the following point:

```
var port = 3000;
app.listen(port);
console.log('Listening on port ' + port);
```

We modify the `package.json` file at this point. We need to include all the Node.js modules that are used by our app:

```
"dependencies": {
        "forecast.io": "latest",
        "util": "latest",
        "express": "latest",
        "jade": "latest"
    }
```

It's time to code the `Interface.jade` file.

We first include jQuery, the Bootstrap CSS file (which is used to give a nice look to our app), and interface the JavaScript file. Then, we define the containers for the different variables:

```
html
  head
    title Weather Station Interface
    script(src='https://code.jquery.com/jquery-2.1.1.min.js')
    script(src='/js/interface.js')
    link(rel='stylesheet', href="https://maxcdn.bootstrapcdn.com/
bootstrap/3.3.0/css/bootstrap.min.css")
  body
    .container
      h1 Weather Station Interface
```

```
h3.row
  .col-md-4
    div Local data
  .col-md-4
    div#temperature Temperature:
  .col-md-4
    div#light Light level:
```

After this, in the JavaScript file, we refresh the containers by calling the API of our app in JavaScript:

```
$.get('/api/temperature', function(json_data) {
    $('#temperature').html('Temperature: ' + json_data.temperature +
' C');
    $.get('/api/light', function(json_data) {
      $('#light').html('Light level: ' + json_data.light + ' %');
    });
});
```

We are now ready to test our application.

> For your convenience, you can download the complete code from the following link:
> https://github.com/marcoschwartz/galileo-blueprints-book

Just as we did in the previous chapter, we also upload the code to the board and build it. Refer to the previous chapter for detailed instructions on this step.

You have to be patient as it takes a while for the Node.js modules to be installed completely. You will be able to see that Express is being installed. You'll know that it is currently being installed as you'll see something similar to this:

```
npm http GET https://registry.npmjs.org/socket.io
npm http GET https://registry.npmjs.org/forecast.io
npm http GET https://registry.npmjs.org/util
npm http GET https://registry.npmjs.org/jade
npm http GET https://registry.npmjs.org/express
npm http 304 https://registry.npmjs.org/socket.io
npm http 304 https://registry.npmjs.org/forecast.io
npm http 304 https://registry.npmjs.org/express
npm http 304 https://registry.npmjs.org/jade
npm http 304 https://registry.npmjs.org/util
```

We will test the API now. Try typing the following line into your browser (of course by changing the IP address with the address of your Galileo board):

```
http://192.168.1.103:3000/api/temperature
```

It should be able to return the temperature reading:

```
{"temperature":30.56640625}
```

Congratulations! You have been able to retrieve the data and display it in your browser.

If you're not getting the anticipated result, you can just go back and check on a couple of things. You can double-check the wirings on your hardware - go back to the previous chapter for the schematic. You can also check the software installed on your board and ensure that all the Node.js modules are installed. To do this, ensure that you have the correct `package.json` file with all the modules required for this project, and that you installed them using the `Build` command of the XDK. Otherwise, you will, for sure, get an error.

Getting the current weather forecast

Once we structure our web application, we will configure Forecast.io. **Forecast.io** is a web API that returns weather forecasts of your exact location, and it updates them by the minute. The API has stunning maps, weather animations, temperature unit options, forecast lines, and more.

We will use this API to integrate global weather measurements with our local measurements.

To do so, first go to the following link:

```
http://forecast.io/
```

Then, look for the **Forecast API** link at the bottom of the page. It should be under the **Developers** section of the footer:

OTHER APPS:	DEVELOPERS:	BLOG
FORECAST.IO	FORECAST API	CONTACT
DARK SKY FOR iOS	OUR DATA SOURCES	PRIVACY
3RD PARTY APPS	HTML EMBED / WIDGET	ADVERTISE

You need to create your own Forecast.io account. You can do so by clicking on the **Register** button on the top right-hand side portion of the page. It will take you to the registration page where you need to provide an e-mail address and a strong password.

Then, you will be required to get an API key, which will be displayed in the Forecast. io interface:

Write this down somewhere, as you will need it soon.

We will then use a Node.js module to get the forecast. The steps are described in more detail at the following link:

`https://github.com/mateodelnorte/forecast.io`

Next, you need to determine the latitude and longitude that you are currently in. Head on to the following link to generate it automatically:

`http://www.ip2location.com/`

Then, we modify the `main.js` file:

```
var Forecast = require('forecast.io');
var util = require('util');
```

This will set our API key with the one used/returned before:

```
var options = {
  APIKey: 'your_api_key'
},
forecast = new Forecast(options);
```

Next, we'll define a new API route for the forecast. This is also where you need to put your longitude and latitude:

```
app.get('/api/forecast', function(req, res) {
  forecast.get('latitude', 'longitude', function (err, result, data) {
    if (err) throw err;
    console.log('data: ' + util.inspect(data));
    res.json(data);
  });
});
```

We will also modify the `Interface.jade` file with a new container:

```
h3.row
  .col-md-4
    div Forecast
  .col-md-4
    div#summary Summary
```

In the JavaScript file, we will refresh the field in the interface.

We simply get the summary of the current weather conditions:

```
$.get('/api/forecast', function(json_data) {
    $('#summary').html('Summary: ' + json_data.currently.summary);
});
```

Again, the complete code can be found at the following link:

```
https://github.com/marcoschwartz/galileo-blueprints-book
```

After downloading, you can now build and upload the application to the board.

Next comes the fun part, as we will test our creation.

Go to the IP address of your board with port `3000`:

```
http://192.168.1.103:3000/
```

You will be able to see the interface as follows:

Weather Station Interface		
Local data	Temperature: 30.078125 C	Light level: 68.359375 %
Forecast	Summary: Partly Cloudy	

Congratulations! You have been able to retrieve the data from Forecast.io and display it in your browser.

If you're not getting the expected result, don't worry. You can go back and check everything. Ensure that you have downloaded and installed the correct software on your board. Also ensure that you correctly entered your API key in the application.

Of course, you can modify your own interface as you wish. You can also add more fields from the answer response of Forecast.io. For instance, you can add a Fahrenheit measurement counterpart. Alternately, you can even add forecasts for the next hour and for the next 24 hours.

Just ensure that you check the Forecast.io documentation for all the fields that you wish to use.

Posting data on Twitter

Finally, you will learn how to post the measured data on a Twitter account.

The first step is to go to the Twitter apps website:

```
https://apps.twitter.com/
```

Then, click on the **Create New App** button.

Fill up the form with the details regarding this app. You can copy the details that I've used in my form, which you can see in the following screenshot:

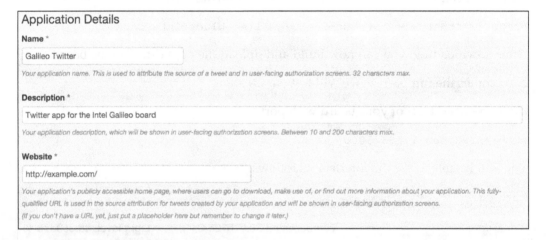

After filling this up, the next thing that you should do is to get the keys and the token. These additional details are required for the authentication of requests on behalf of Twitter. These are considered as passwords, so ensure that you do not share these things to untrusted people.

You should be able to see the interface as shown here:

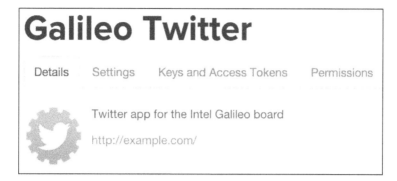

Next, you will be required to set the permissions. This decision depends on the needs of your application. In my case, I have set it to **Read and Write**. I deem that accessing direct messages is not necessary for our experiment.

You will be able to see the interface as follows:

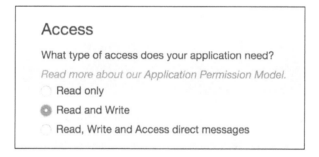

All the things that we needed to do on Twitter are now done, so let's go back to editing main.js to integrate the Twitter API into the application.

We will be using the following twit module:

```
var Twit = require('twit');
```

We need to enter our Twitter credentials here:

```
var T = new Twit({
    consumer_key:         '..'
  , consumer_secret:      '..'
  , access_token:         '..'
  , access_token_secret:  '..'
});
```

Next, we will define a function that can measure data, format it, and tweet it:

```
function tweetData() {
    // Measurements
    var b = temp_sensor_pin.read();
    var temperature = (b/4096*5000 - 500) / 10;

    var a = light_sensor_pin.read();
    var light_level = a/4096*100;

    // Message
    var message = 'Temperature is ' +
    temperature + ' C and light level is ' +
    light_level + ' %.';

    // Tweet
    T.post('statuses/update', { status: message }, function(err, data,
response) {
        if (err) {console.log(err)};
    });
}
```

Finally, we will set this function to execute every minute:

```
tweetData();
setInterval(tweetData, 60000);
```

You can go to the following link to download the complete code used in this project:

```
https://github.com/marcoschwartz/galileo-blueprints-book
```

As usual, upload the code, build it, and run it on your board.

You are now ready to check your Twitter account. A tweet containing the measurement readings that you have specified should be visible now. If you can wait for the next minute, it should tweet the same thing again.

As an example, here is a snapshot of my tweet, on an account I created just for this project:

 Open Source Home @opensourcehome · 12s
Temperature is 30.56640625 C and light level is 70.99609375 %.

Congratulations! You have been able to make the Galileo board interact with web APIs. There's a wide variety of applications that you can perform with these principles.

However, if your tweet is not getting through, you can go back and check on a couple of things. Check whether the correct software is downloaded and installed on your board. Check your Twitter details, keys, and tokens. Remember that the input on the software is case-sensitive in the code. You may have just missed a character.

Another piece of good news is that you can control the tweets. You can change the settings such as the tweet interval, and you can tweet other messages. You can create alerts for certain situations, for example, if the temperature is getting too low.

Summary

To make our Galileo board get the current weather measurements and broadcast it to Twitter, we have performed the following steps:

First, we organized and structured our web app with Express. Node.js was responsible for the data and measurement, while Jade was used to build the interface. We also use jQuery to handle the interface, and Bootstrap CSS to give a better look to the app.

Then, we learned to add an online forecast to the app. Finally, we learned how to interface our app with Twitter so we could tweet the data on our twitter accounts.

In the next chapter, we will continue with a more complex Internet of Things application. We will make the Galileo board into an online data-logging platform, so that we can access and plot the data from any web browser.

6
Internet of Things with Intel Galileo

In the last chapter, we learned how to interact with web APIs, such as Twitter. We were able to forward the measurements that we got from the sensors to our Twitter feed.

In this chapter, we will take advantage of what we learned in the previous chapter and raise the bar. We will integrate our Galileo board into an **Internet of Things (IoT)** framework.

IoT is an emerging and exciting technology, which is the interconnection of computers, mobile phones, tablets, and any computing device, to the Internet infrastructure. IoT fosters automation, faster cloud computing, supply chain connection between consumers and partners, and a whole lot more.

In the project completed in the previous chapter, we visualized the data locally. This time, we will do it on Cloud services.

In completing this project, there will be three major steps involved:

1. First, we need to log data in the Cloud using **Dweet.io**.
2. Then, we will give structure to the data and monitor it using **Freeboard.io**.
3. Finally, we will store data on a Google Docs spreadsheet via **Temboo**.

Let's start!

Logging your data in the cloud

As discussed, we will first log the data in the cloud. Here, we will use the same hardware and software configurations that we used in the previous chapter.

To log data in the cloud, we will use a service called Dweet.io. Dweet.io is a simple publishing and subscribing platform created for machines, gadgets, and any form of computer. There's no sign up or set up required so you can use it instantly.

 Here is the link to Dweet.io:
http://dweet.io/

We will use the following code. Let's examine it first so that you can understand it easily:

1. We create a new XDK project first.

2. Then, we include the following libraries, which are used to communicate with the pins of the Galileo board:

```
var mraa = require("mraa");
var util = require('util');
var request = require('request');
```

3. We need to set the analog pins and set their resolution to 12 bits:

```
var light_sensor_pin = new mraa.Aio(0);
light_sensor_pin.setBit(12);
var temp_sensor_pin = new mraa.Aio(1);
temp_sensor_pin.setBit(12);
```

4. We then define the function send_data().

5. The first thing that we will measure is the light level. Here, we assign the sensor to read it and we compute and convert the data to obtain a percentage result:

```
var a = light_sensor_pin.read();
var light_level = a/4096*100;
light_level = light_level.toPrecision(4);
console.log("Light level: " + light_level + " %");
```

We will do the same for the temperature measurement. We assign another sensor to measure it.

6. Then, we will add equations so that the displayed temperature result will be in degree Celsius:

```
var b = temp_sensor_pin.read();
var temperature = (b/4096*5000 - 500) / 10;
temperature = temperature.toPrecision(4);
console.log("Temperature: " + temperature + " C");
```

7. You should give a unique name to your device so you can easily identify it. You can use your own name for instance.

```
var device_name = 'galileo_5etr6b';
```

8. Next, we will build the url:

```
var device_name = 'galileo_5etr6b';
var dweet_url = 'https://dweet.io/dweet/for/' + device_name +
'?temperature=' + temperature + '&light=' + light_level;
console.log(dweet_url);
```

9. We will also set an option to request data from Dweet.io:

```
var options = {
  url: dweet_url,
  json: true
};
```

10. Then, we will send that request to Dweet.io:

```
request(options, function (error, response, body) {
  if (error) {console.log(error);}
  console.log(body);
});
```

11. We will set our request to be sent every 10 seconds for us to be able to get the real-time data:

```
send_data();
setInterval(send_data, 10000);
```

12. Don't forget to modify `package.json` to include all the Node.js modules that are used in this project:

```
{
  "name": "Dashboard",
  "description": "",
  "version": "0.0.0",
  "main": "main.js",
  "engines": {
```

```
        "node": ">=0.10.0"
    },
    "dependencies": {
        "util": "latest",
        "request": "latest"
    }
}
```

 Note that you can download the complete code from the following GitHub link:

https://github.com/marcoschwartz/galileo-blueprints-book

13. Now, we will build the code, upload it to our Galileo board, and run it. If all goes well, you should be able to receive an answer every 10 seconds. Here is an example of what you will see in your console:

```
https://dweet.io/dweet/for/galileo_5etr6b?temperature=25.1953125&light=40.625
{ this: 'succeeded',
  by: 'dweeting',
  the: 'dweet',
  with:
   { thing: 'galileo_5etr6b',
     created: '2014-12-23T09:03:41.363Z',
     content: { temperature: 25.1953125, light: 40.625 } } }
Light level: 40.72265625 %
Temperature: 25.1953125 C
https://dweet.io/dweet/for/galileo_5etr6b?temperature=25.1953125&light=40.72265625
{ this: 'succeeded',
  by: 'dweeting',
  the: 'dweet',
  with:
   { thing: 'galileo_5etr6b',
     created: '2014-12-23T09:03:51.440Z',
     content: { temperature: 25.1953125, light: 40.72265625 } } }
```

14. After this, check results in Dweet.io by visiting the following link:

https://dweet.io/get/latest/dweet/for/my-thing-name

15. Don't forget to change "my-thing-name" to the name of your own device. Otherwise, you won't get anything.

Here is an example of a response that you will see in your browser:

{"this":"succeeded","by":"getting","the":"dweets","with":[{"thing":"galileo_5etr6b","created":"2014-12-23T09:04:11.486Z","content":{"temperature":25.1953125,"light":40.4296875}}]}

Congratulations! Your data is constantly being logged to Dweet.io.

If you have problems, you can always go back and check on a few things:

- First, double check your hardware connections. You can refer to the previous chapter for this.

- Next, ensure that the code you have uploaded is the correct one.

- Finally, check the code, specifically for the unique name that you have used for your own device.

Monitoring your data in the cloud

Now, we will visualize this data on the dashboard using Freeboard.io.

Again, Freeboard.io is an open source and production-ready platform, which can provide a dashboard for your data. The data that we have sent using Dweet.io can be integrated to Freeboard.io, so we will have easy-to-understand visuals of the data.

Just visit the following link to access Freeboard.io:
`http://freeboard.io`

After this, you will need to perform the following steps:

1. First we create an account to be able to use it. However, if you have already created one, you can just log in.

2. Then, we create a new board by giving it a name:

I named mine 'Galileo'. You are free to choose yours:

3. Next, we will click on **Add** under the **Datasources** tab:

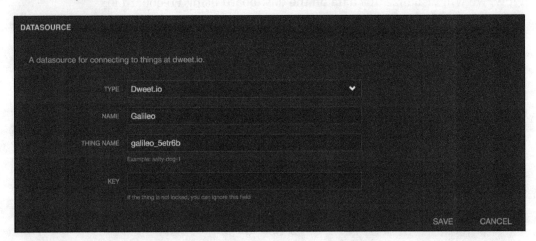

4. We will configure our data source here. Select **Dweet.io** and type in the name of your device:

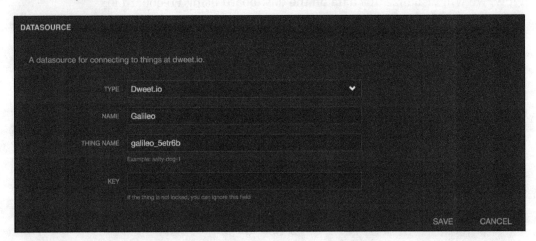

5. We now have our data source. We can see its name and the timestamp of its last update. You can click on the **refresh** button to manually refresh the data:

6. We will then add a new pane by clicking on the **Add pane**:

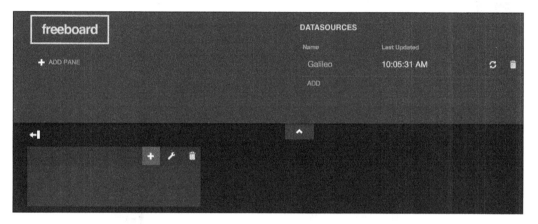

7. Then, we create a new widget. A new page with the widget details will pop up.

8. You need to provide the necessary details. We will use a **Gauge** type of widget and give it a title.

   In this project, we only have two types of data — light level and temperature. This example is for the temperature display.

   We link the display to the temperature datasource. Refer to what I have typed in the **Value** box. Change the name **Galileo** to the name that you have given to your own device.

Specify the temperature measurement details, such as the unit, minimum reading, and maximum reading. In my case, I have used **C**, **0** and **40**, respectively:

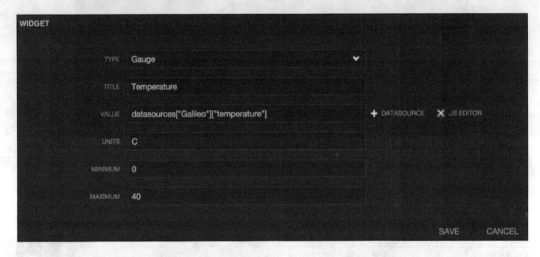

9. Click on **SAVE**. You will see that the temperature gauge is displayed and refreshed every 10 seconds:

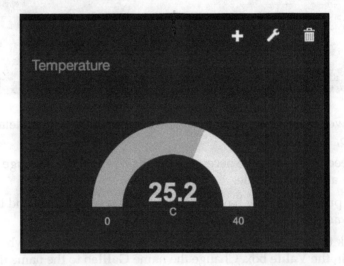

10. Create a new gauge widget and do the same for light level measurements:

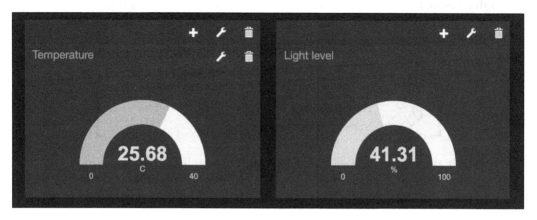

Way to go! You now have a dashboard that you can access from just about anywhere.

If your display is not working, you can look back at the steps that we performed. Check the data that you have typed in- you might just have missed something small. Ensure that you are specifying the correct name for your device.

Storing and plotting your data in the cloud

At this point, we are ready to take on the third major step, which is to store and plot our data in the cloud. In this project, we will use a Google Docs spreadsheet to store the data. It's free and easy to use, and can also be accessed from anywhere as long as you have an internet connection.

The great thing about this is that you can plot your data live. This means you will be able to see a graph of the real-time data.

You will just need a Gmail account to use Google Docs.

Temboo will be responsible for writing the data to our spreadsheet.

Temboo is an Internet of Things (IoT) platform, which can be used to program devices, code utilities, databases, languages, APIs, and a lot more. It can be used, for example, to store data online, send automated text messages, or access social networks.

Temboo is ready for use with **MicroControllers Units** (**MCUs**) and mobile apps through an SDK library or Galileo SDKs. It can connect to devices through Bluetooth, Ethernet, GSM, and Wi-Fi. For this, you need to perform the following steps:

1. The first thing to do is go to Temboo's official website and create an account with them:

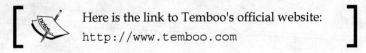

   Here is the link to Temboo's official website:
   `http://www.temboo.com`

2. After signing up, you can create an app. You will get a name, an app name, and an app secret key. Write these down. You will need them in the following steps. Don't share this information with people you don't trust. These are similar to passwords so this is sensitive information.

3. Next, go the the page of the library called `AppendRow`, which is available at

   `https://www.temboo.com/library/Library/Google/Spreadsheets/`
   `AppendRow/`

   You will be presented with something like this:

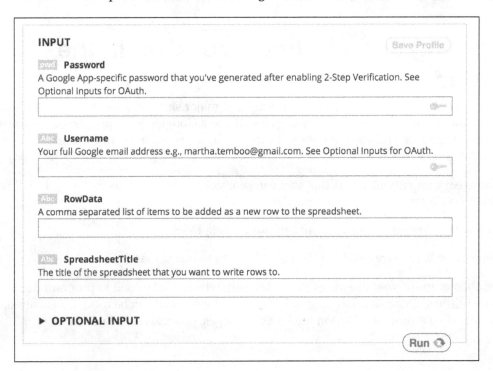

1. The first thing you need is the application-specific password for Gmail. This will let Gmail allow Temboo to access your Google Docs. To get this, go to the following link:

   ```
   https://security.google.com/settings/security/
   apppasswords
   ```

2. Once you have retrieved your password, enter your credentials on the page. Decide the username, row data, and spread sheet title. When you have finished filling up the form, you can test it by clicking on **Run**.

3. The code will be generated automatically:

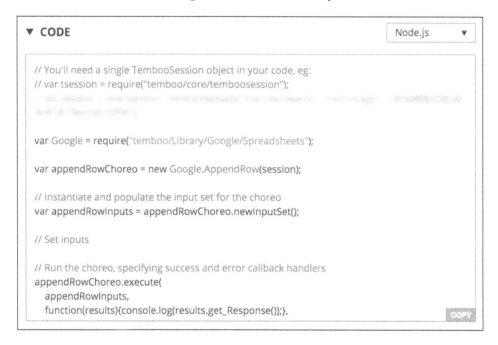

```
▼ CODE                                              Node.js    ▼

// You'll need a single TembooSession object in your code, eg:
// var tsession = require("temboo/core/temboosession");


var Google = require("temboo/Library/Google/Spreadsheets");

var appendRowChoreo = new Google.AppendRow(session);

// Instantiate and populate the input set for the choreo
var appendRowInputs = appendRowChoreo.newInputSet();

// Set inputs

// Run the choreo, specifying success and error callback handlers
appendRowChoreo.execute(
  appendRowInputs,
  function(results){console.log(results.get_Response());},         COPY
```

Besides this, you can also find the code in the Github repository of this book.

4. Once you have the code, go to Google Drive:

   ```
   https://drive.google.com/
   ```

5. After this, create a new spreadsheet and give it a name.

6. Then, give a title to the columns. You can refer to the following screenshot for the titles I have used:

	A	B	C
1	**Date**	**Temperature**	**Light level**
2			
3			
4			
5			
6			
7			
8			
9			
10			

We will look at the details of this code in XDK:

1. We first include the libraries:

```
var mraa = require("mraa");
var util = require('util');
```

2. Then, we define the pin for the light sensor and the temperature sensor. We have also set the data bit for these pins:

```
var light_sensor_pin = new mraa.Aio(0);
light_sensor_pin.setBit(12);
var temp_sensor_pin = new mraa.Aio(1);
temp_sensor_pin.setBit(12);
```

3. We then create a Temboo session by filling in our credentials:

```
var tsession = require("temboo/core/temboosession");
var session = new tsession.TembooSession("your_temboo_name",
"your_temboo_app", "your_temboo_key");
```

4. Next, we include the `temboo` library that we will need, which is Google Spreadsheets:

```
var Google = require("temboo/Library/Google/Spreadsheets");
```

We then define the `send_data()` function again.

5. We next fix the measure light level code. This code reads the data from the pin and converts the data in such a way that we will end up with a percentage light level measurement. We need to divide the reading by 4096 here, as this is the total number of levels that can take the analog input on the Galileo (12-bits Analog-Digital Converter):

```
var a = light_sensor_pin.read();
var light_level = a/4096*100;
light_level = light_level.toPrecision(4);
console.log("Light level: " + light_level + " %");
```

6. We then do the same for the temperature measurements. We read the data from the other pin and add equations so that we get readings in Celsius:

```
var b = temp_sensor_pin.read();
var temperature = (b/4096*5000 - 500) / 10;
temperature = temperature.toPrecision(4);
console.log("Temperature: " + temperature + " C");
```

7. The next thing to do is to create a Temboo object:

```
var appendRowChoreo = new Google.AppendRow(session);
```

8. Then, we create a new input:

```
var appendRowInputs = appendRowChoreo.newInputSet();
```

We will get the date:

```
var d = new Date();
```

We need to input our Gmail account details and create entries for the spreadsheet.

9. Here is the code. Just replace the username and password with your own credentials:

```
appendRowInputs.set_Password("your_gmail_app_password");
appendRowInputs.set_Username("your_gmail_account_name");
appendRowInputs.set_RowData(d + "," + temperature + "," + light_
level);
appendRowInputs.set_SpreadsheetTitle("Galileo");
```

10. We now execute `choreo`:

```
appendRowChoreo.execute(
  appendRowInputs,
  function(results){console.log(results.get_Response());},
  function(error){console.log(error.type); console.log(error.
message);}
);
```

11. We need to send data every 10 seconds:

```
send_data();
setInterval(send_data, 10000);
```

Note that the 10-second delay should be used for demo purposes only. Temboo is limited by the number of calls made.

To adapt for the limited number of free calls, you can schedule the measurement to occur every hour or so, just for normal everyday use.

12. Don't forget to modify the `package.json` file:

```
{
  "name": "Dashboard",
  "description": "",
  "version": "0.0.0",
  "main": "main.js",
  "engines": {
    "node": ">=0.10.0"
  },
  "dependencies": {
      "util": "latest",
      "temboo": "latest"
  }
}
```

Note that you can download the complete code from the Github repository:

```
https://github.com/marcoschwartz/galileo-
blueprints-book
```

13. Now build, upload, and run the code in the XDK again.

14. Once the code is running, you will get a console log indicating that the data was stored in the spreadsheet:

```
success
Light level: 41.41 %
Temperature: 25.20 C
success
Light level: 41.80 %
Temperature: 25.20 C
success
Light level: 42.29 %
Temperature: 25.20 C
success

Intel XDK — Message Received: stop
=> Stopping App <=
```

If you go over your Google Docs spreadsheet, you will also see the data logged live into the sheet:

	A	B	C
1	Date	Temperature	Light level
2	Tue Dec 23 2014 10:01:11 GMT+0000 (UTC)	25.2	42.19
3	Tue Dec 23 2014 10:01:23 GMT+0000 (UTC)	25.2	41.8
4	Tue Dec 23 2014 10:01:33 GMT+0000 (UTC)	25.2	41.41
5	Tue Dec 23 2014 10:01:43 GMT+0000 (UTC)	25.2	41.8
6	Tue Dec 23 2014 10:01:53 GMT+0000 (UTC)	25.2	42.29

You can also use Google Docs to plot the data coming in:

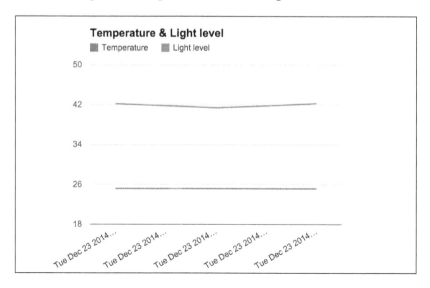

These plots will also be updated in real time, so you can monitor the state of the sensor from anywhere in the world in real time!

Great! You now have a way to store your data online and plot the data live on a spreadsheet.

If you have trouble with making the data appear on the spreadsheet, you can take a look at the steps again. Ensure that you have downloaded and built the correct code. Check on your credentials- you may just have a typo error. Good luck!

Summary

In this chapter, you learned how to send measured data to online services and monitor it online. We have immersed ourselves into the IoT framework. With the free services and basic principles that you have learned here, you can construct many more powerful projects.

We accomplished three major steps in this project. First, we exported data through Dweet.io. Then, we integrated the data into Freeboard.io to create a dashboard for our data. Next, we tapped on Temboo to input and plot the data on a Google Docs spreadsheet.

In the next chapter, you will learn how to control your Galileo board from anywhere in the world. Imagine switching off an appliance in your home while you're on vacation in Europe or Asia! That would be really handy!

7
Controlling Your Galileo Projects from Anywhere

In the previous chapter, we integrated our Galileo board with an IoT framework. We measured data on the board and sent the data through online services.

Furthermore, we enabled the data to be stored and visualized online through the cloud.

In this chapter, you will learn how to develop the software that will control your board from anywhere. For instance, you will learn how to switch the state of the relay from a remote browser.

Here is an overview of how we will do this:

- Plug a relay to the board
- Make a local interface to control this relay
- Set up our system so that we can access it from anywhere

We will discuss several ways to do this, so you can have an overview of what works best. We will use a little software running on our computer to relay the information from the web to the Galileo board. We will also see how to modify our router settings to do the same without the use of a computer.

Hardware and software requirements

We won't have trouble with this project's requirements as we will use the same hardware and software that we used in the previous chapter. The details about the hardware requirements and configurations are given in *Chapter 4, Monitoring Data Remotely*.

All we need to do is to add a 5V relay on the board.

 For more information about the relay, you can refer to its product link:
Pololu 5V relay module (`http://www.pololu.com/product/2480`)

The **Pololu relay** is a **single-pole double-throw** (**SPDT**) relay and is rated up to 10A under most conditions. SPDT relays are commonly used in electronics. They have two input pins for the coil, and three output pins to control electrical devices. SPDT utilizes a Zener diode for fast current decay. It also has two LED lights as coil actuation indicators.

Hardware configuration

Let's take a look at the hardware configuration just to refresh our memory.

Here is the schematic used for this project:

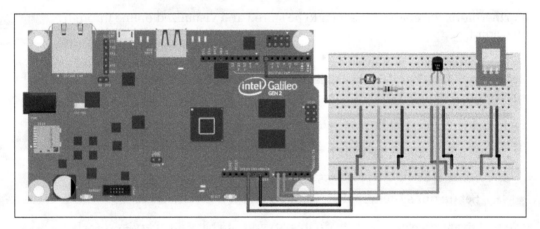

We used the same configuration in *Chapter 3, Controlling Outputs Using the Galileo Board*. This is the relay that we will use in this chapter:

If you want to go back to the specific instructions as to how to do this, you can go back to *Chapter 3, Controlling Outputs Using the Galileo Board*.

The following picture will show you my final configuration:

Building an interface to control the relay

The next thing that we need to do is to build an interface to control the relay.

The **On** and **Off** buttons will serve as our trigger controls.

Just a note: the code that we will use here is similar to the code that we used in *Chapter 5, Interacting with Web APIs*. Nevertheless, we will still go over the code that we will be using:

1. First, we define the relay pin object:

   ```
   var relay_pin = new mraa.Gpio(7);
   relay_pin.dir(mraa.DIR_OUT);
   ```

2. Then, we will use the same API we built in the previous chapter, and add functionalities to it. Create a new API call for the relay using the following code:

   ```
   app.get('/api/relay', function(req,res){

     // Get desired state
     var state = req.query.state;
     console.log(state);

     // Apply state
     relay_pin.write(parseInt(state));

     // Send answer
     json_answer = {};
     json_answer.message = "OK";
     res.json(json_answer);
   });
   ```

3. In this code, we determine the state of the relay from the query. Then, we will write this state onto the pin. Finally, we send an OK JSON answer.

4. After this, we modify the interface defined in the file called `interface.jade`, which is located in the views folder. The goal is to have two buttons for the two states of the relay, on and off:

   ```
   h3.row
       .col-md-4
       div Relay control
       .col-md-4
         button.btn.btn-lg.btn-block.btn-primary#on On
       .col-md-4
         button.btn.btn-lg.btn-block.btn-danger#off Off
   ```

5. We will also modify the JavaScript file in order to link the buttons to the API:

```
$('#on').click(function() {
  $.get('/api/relay?state=1');
});


  ('#off').click(function() {
  $.get('/api/relay?state=0');
});
```

We are now ready to test the project on the Galileo board.

 For your convenience, you can download the complete code from the Github link:

`https://github.com/marcoschwartz/galileo-blueprints-book`

6. Just as what we did in the previous chapters, once the download of the code is complete, we will upload the code to the Galileo board and build it.

7. When finished, we will test the API, and type the following line into our browser:

(Of course, you have to replace the IP address with the address of your own Galileo board.)

`http://192.168.1.103:3000/api/relay&state=1`

It should be able to return the status as follows:

`{"message":"OK"}`

The additional thing, which can tell you whether everything works just fine, is the audible 'clicking' sound of the relay.

8. Now, try typing the following line (again, use your own IP address):

`http://192.168.1.103:3000`

Once you have sent this command, you should be able to see the interface, which will look something like this:

Weather Station Interface		
Local data	Temperature: 29.59 C	Light level: 63.77 %
Relay control	On	Off

Now comes the fun part! Try clicking on the **On** and **Off** buttons. They should be able to control the relay flawlessly.

Congratulations! You have built an interface that can control the relay.

If things are not working as desired, you can check on a couple of things:

1. First, check the hardware configurations. You can refer to *Chapter 3, Controlling Outputs Using the Galileo Board* for details.

2. Next, ensure that you have uploaded the correct code in to your Galileo board.

3. Another thing that can mess things up is the IP address that you have entered. Ensure that it is indeed the IP address of your board. You can get the IP address of the board from the Intel XDK software. Good luck!

Accessing the interface from anywhere

Now that we have successfully controlled the relay through the interface, we will explore ways to access this interface from anywhere.

For this, we will use **Ngrok**. This is a simple utility that enables its users to test local hosts with remote APIs. Through Ngrok, collaborators can overcome firewalls and access TLS connections.

It is an easy-to-use service, which requires no sign ups.

The Ngrok service will open up a tunnel between your local board and the Web. Then, you can access the interface from anywhere, anytime! It should be noted, however, that it doesn't work with the Galileo as of the time of this writing.

So, we will use our own computer as a relay instead, establishing the connection between the Web and your Galileo board:

1. To do so, we first download the Ngrok files from the following URL:

   ```
   https://ngrok.com/download
   ```

2. Now, if you are using Windows, double-click on the file to unzip it. If you are using Linux or OS X, go to the folder where you have downloaded the files and unzip the file by typing the following:

   ```
   unzip ngrok.zip
   ```

3. Next, type the following:

   ```
   ./ngrok 192.168.1.103:3000
   ```

4. Again, replace the IP address with your own IP address; otherwise, it won't work.

5. After typing the last line, you should see the tunnel being established:

```
ngrok                                                    (Ctrl+C to quit)

Tunnel Status         online
Version               1.7/1.7
Forwarding            http://4681858b.ngrok.com -> 192.168.1.103:3000
Forwarding            https://4681858b.ngrok.com -> 192.168.1.103:3000
Web Interface         127.0.0.1:4040
# Conn                0
Avg Conn Time         0.00ms
```

6. The next thing to do is to copy this URL and enter it in your browser. You will be presented with the same interface as earlier:

7. Check whether the buttons are working. Click on the **On** button and check the relay. Do the same for the **Off** button.

Great! You can now access the interface even when you are outside your network. This means that you can now access your board from anywhere. Good job!

If the Ngrok service doesn't work out for you, you need to check a few things:

* First, ensure that you properly downloaded the Ngrok files, and that you are running the Ngrok executable from the right folder.

* Another thing that you can check is whether you have entered the right IP address of your own Galileo Board. Good luck!

Removing the need for a third party

In the last section, we saw how to access our interface from anywhere in the world by running software on our computer, called Ngrok. However, we will see that there are other ways to do the same without using Ngrok: but these other ways are a bit more complicated.

One way is to use the parameters of your internet router to automatically forward the connections from outside to your local Galileo board. For this, you need to perform the following steps:

1. First, you need to know the IP address of your router.

    [Note that this is not the IP address of your local computer, but the one of your router. Visit the following link:

    http://www.whatismyip.com/]

    Your IP address will immediately be printed on the screen:

    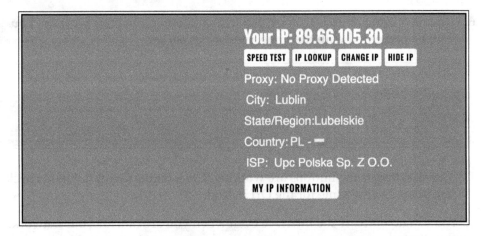

2. Now, we need to modify our router settings to forward connections from this external IP to our Galileo board. We'll use a given port for this, such as 3001.

3. First, you need to know the local IP address of your router. This can be done in most systems by looking into the **Network Preferences**. On Windows, it is called **Network and Sharing Center** For example, inside the **Network Preferences** of OS X:

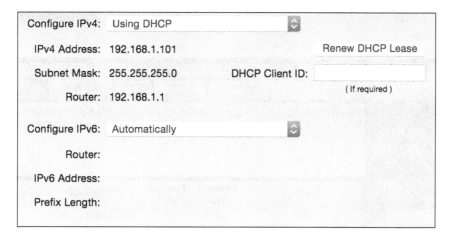

4. What we are looking for here is the **Router** IP address. Now, go to this address inside your favorite web browser. Depending on your router, you will be taken to the login interface:

5. Enter your **User Name** and **Password** now. If you don't remember them, you can usually find them inside the manual of your router.

6. Then, we come to the next step, which depends on the router you have. You need to look for a page called **Forwarding** or **Port forwarding**, and on this page, you need to set the external port on which you will access the Galileo board, for example, 3000.

7. Then, you need to check the **TCP** as this is the protocol we'll use.

8. Finally, enter the local IP address of your Galileo board. Now save everything, and reboot the router.

Here is a screenshot of the port forwarding on the router menu that I used for this tutorial:

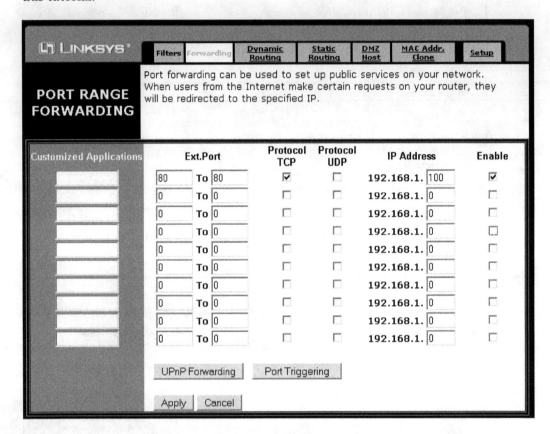

You can now access the board again via the port you defined earlier. For example, with the IP address I defined earlier, I can go to the following URL in my browser:

```
http://89.66.105.30:3000
```

I will then have access to the exact same interface as used earlier. Also note that this IP address might change over time as most Internet Service Providers use a dynamic DNS. Therefore, you might need to repeat the procedure from time to time.

Summary

In this chapter, you learned how to control a device such as a relay remotely, from anywhere in the world.

Here is a summary of what we did:

First, we attached a relay to our board just as we did in *Chapter 3, Controlling Outputs Using the Galileo Board*. Then, we built an interface to control the relay via simple buttons. Finally, we integrated the interface through Ngrok, which enabled us to access the controls from any place where there is an internet connection. We also looked at another way to access the interface, which is to modify the router settings and forward the incoming connections to the router.

In the next chapter, we are going to build a wireless security camera just using a USB webcam. Stay tuned!

8
Displaying the Number of Unread Gmail E-mails on an LCD Screen

In the previous chapter, we developed software that allowed us to control our Galileo board from anywhere.

In this chapter, you will learn how to integrate another existing application into our Galileo board. We will use its onboard Ethernet connection to connect to our Gmail account. We won't need the help of our computer in doing this: we will simply display the e-mail data on an external LCD screen using the Galileo board.

In this project, we will get the number of unread e-mails in your Gmail account, and display this count on the LCD screen.

Let's start!

Hardware and software requirements

To complete this project, we will need only two major components. First is the Galileo board, of course, and the second one is the LCD screen.

The board that I've used is, again, the Intel Galileo Gen 2 board.

For the LCD screen, I've used the Saint Smart LCD. It's an IIC/I2C/TWI 20x4 LCD with a contrast control knob selector and a backlight. If you don't have this specific type of LCD, you can use any other I2C LCD: it won't be a problem.

You will also need a MicroSD card to store your data.

Here is the picture of the Saint Smart LCD that I have used in this project:

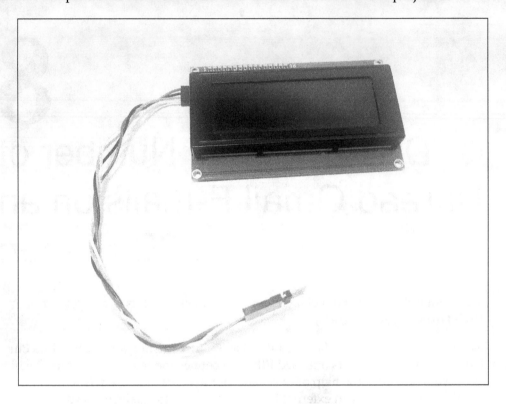

Here is the list of all the components used in this project:

- Intel Galileo Gen 2 board at `https://www.sparkfun.com/products/13096`
- I2C LCD screen at `http://www.sainsmart.com/sainsmart-iic-i2c-twi-serial-2004-20x4-lcd-module-shield-for-arduino-uno-mega-r3.html`
- MicroSD card

On the software side, you will need the library for the LCD screen you are using. Note that the library depends on the manufacturer of the LCD screen. For the screen that I have used in this project, you can find the library at the following link:

`https://github.com/fdebrabander/Arduino-LiquidCrystal-I2C-library`

To install this Arduino library, simply place the whole library folder inside the `/libraries` folder of your main Arduino installation folder.

Hardware configuration

When you get everything you need in place, you may start configuring the hardware. The configuration is pretty simple.

Start by connecting the LCD screen to the Galileo board. To do so, connect the VCC pin of the LCD to the 5V port of the board. Then, connect the LCD's GND to the board's GND pin. Connect the SDA and SCL pins accordingly. You can see the board's SDA and SCL pins labeled as such.

Here is a picture of how the final configuration looks:

Once the configuration is finished, insert the MicroSD card into the Galileo board and connect it to the Ethernet.

Testing the LCD screen

We are now ready to test the LCD screen. The following is the complete code that I've used to do so:

```
// Libraries
#include <Wire.h>
#include <LiquidCrystal_I2C.h>
#include <SD.h>
#include <Ethernet.h>

// LCD Screen
LiquidCrystal_I2C lcd(0x27,20,4);

void setup()
{
  // Initialize LCD screen
  initDisplay();

}

void loop()
{
  // Print message on the LCD screen
  lcd.setCursor(0,0);
  lcd.println("This is a test of the LCD screen!");
}


// Initialize LCD screen
void initDisplay()
{
  lcd.init();
  lcd.backlight();
  lcd.clear();
}
```

Let's look at the code line by line:

The first part of the code includes the libraries that we need in order to carry out this project. The four libraries that you should include are Wire.h, LiquidCrystal_I2C.h, SD.h, and Ethernet.h:

```
#include <Wire.h>
#include <LiquidCrystal_I2C.h>
#include <SD.h>
#include <Ethernet.h>
```

Then, we declare the LCD instance. If you have used a different screen size as I have, you can just modify the attributes: the number of lines and characters of the screen.

First, change the numbers to fit your screen details:

```
LiquidCrystal_I2C lcd(0x27,20,4);
```

This is just an init display in the setup() function:

```
initDisplay();
```

These codes will print a simple message in the loop() function. You can modify the message if you like:

```
lcd.setCursor(0,0);
lcd.println("This is a test of the LCD screen!");
```

What follows are the details of the init() function:

```
void initDisplay()
{
  lcd.init();
  lcd.backlight();
  lcd.clear();
}
```

For your convenience, you can download the complete code from the following link:

https://github.com/marcoschwartz/galileo-blueprints-book

Before we're ready to test, there's one more thing that we should do: we should upload the code using the Arduino IDE for the Galileo board.

If everything works well, you will see messages being printed on the LCD screen.

If there are no messages being printed, then check your hardware connections and the code. You might just have missed a single connection or you might just have missed a character or two in your code.

Printing the unread e-mails on the LCD screen

We can now print the number of unread e-mails on the LCD screen.

The code for this function is composed of two parts:

- The Python script to connect to Gmail
- The code to read the number of unread e-mails that you have in Gmail

We'll also need the Arduino sketch to get this number and display it on the LCD screen.

Let's look at the complete Python script first:

```python
# Import library
import imaplib

# Connect to Gmail
obj = imaplib.IMAP4_SSL('imap.gmail.com', '993')

# Login to Gmail
obj.login('your_gmail_address','your_gmail_password')

# Print number of unread emails
print len(obj.search(None,'UnSeen')[1][0].split())
```

Now, let's analyze it. The first thing that this code does is import the `imaplib` library:

```python
import imaplib
```

Then, it connects to Gmail:

```python
obj = imaplib.IMAP4_SSL('imap.gmail.com', '993')
```

After this, it will enter the Gmail e-mail ID and password. You should replace the dummy characters with your own credentials:

```python
obj.login('your_gmail_address','your_gmail_password')
```

Finally, it will print the number of unread e-mails:

```python
print len(obj.search(None,'UnSeen')[1][0].split())
```

If you have two-factor authentication activated in your account, you need to get an application-specific password.

You can learn how to do this through the tutorial available at the following link:

`https://security.google.com/settings/security/apppasswords?pli=1`

This will bring you to a page where you will require a password just for your app. You will be assigned a password that you need to enter in the Python script, as you can see in the following screenshot:

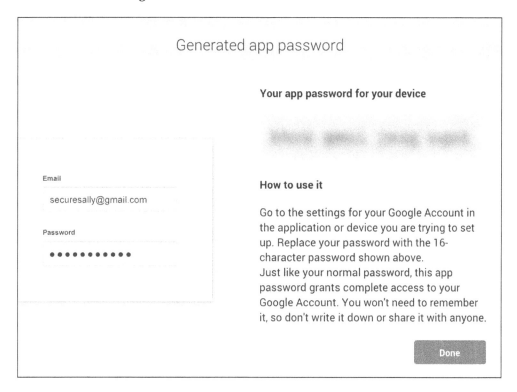

Now, we will see the Arduino code for this part. Here is the complete code:

```
// Libraries
#include <Wire.h>
#include <LiquidCrystal_I2C.h>
#include <SD.h>
#include <Ethernet.h>

// LCD Screen
```

```
LiquidCrystal_I2C lcd(0x27,20,4);

// Ethernet MAC
byte mac[] = {0x98, 0x4F, 0xEE, 0x01, 0x90, 0xC5};

// Email-Checking variables
const int emailUpdateRate = 10000; // Update rate in ms (10 s)
long emailLastMillis = 0; // Stores our last email check time
int emailCount = 0; // Stores the last email count

void setup()
{

    // Serial monitor is used for debug
    Serial.begin(115200);

    // Init SD card
    initSDCard();

    // Init LCD
    initDisplay();

    // Connect Ethernet
    initEthernet()
}

void loop()
{
    // Only check email if emailUpdateRate ms have passed
    if (millis() > emailLastMillis + emailUpdateRate)
    {
        emailLastMillis = millis(); // update emailLastMillis

        // Get unread email count, and print it on the LCD
        int tempCount = getEmailCount();
        lcd.print("You have ");
        lcd.print(tempCount);
        lcd.println(" unread mail.");

        if (tempCount != emailCount) // If it's a new value, update
        { // Do this to prevent blinking the same #
            emailCount = tempCount; // update emailCount variable
            printEmailCount(emailCount); // print the unread count
        }
```

```
  }

  // Bit of protection in case millis overruns:
  if (millis() < emailLastMillis)
    emailLastMillis = 0;
}

// Get email count by executing the Python script
int getEmailCount()
{
  int digits = 0;
  int temp[10];
  int emailCnt = 0;

  // Send a system call to run our python script and route the
  // output of the script to a file.
  system("python /home/root/emails.py > /media/card/emails.txt");

  File emailsFile = SD.open("emails"); // open emails for reading

  // read from emails until we hit the end or a newline
  while ((emailsFile.peek() != '\n') && (emailsFile.available()))
    temp[digits++] = emailsFile.read() - 48; // Convert from ASCII to
a number

  emailsFile.close(); // close the emails file
  system("rm /media/realroot/emails"); // remove the emails file

  // Turn the inidividual digits into a single integer:
  for (int i=0; i<digits; i++)
    emailCnt += temp[i] * pow(10, digits - 1 - i);

  return emailCnt;
}

// Init LCD display
void initDisplay()
{
  lcd.init();
  lcd.backlight();
  lcd.clear();
}

// Init SD card
```

```
void initSDCard()
{
  SD.begin();
}

// Init Ethernet connection
uint8_t initEthernet()
{
  if (Ethernet.begin(mac) == 0)
    return 0;
  else
    return 1;
}
```

Let's look, in detail, at the elements of the code that we have added.

The following line of code will ask you to input your MAC address:

```
byte mac[] = {0x98, 0x4F, 0xEE, 0x01, 0x90, 0xC5};
```

Here, we just declare some variables:

```
const int emailUpdateRate = 10000; // Update rate in ms (10 s)
long emailLastMillis = 0; // Stores our last email check time
int emailCount = 0; // Stores the last email count
```

Then, we start the serial:

```
Serial.begin(115200);
```

Next, we initiate the MicroSD card:

```
initSDCard();
```

And we do the same for the Ethernet:

```
initEthernet()
```

Here is the main loop responsible for the e-mail checks, and executes the Python script and prints data on the LCD:

```
if (millis() > emailLastMillis + emailUpdateRate)
{
  emailLastMillis = millis(); // update emailLastMillis

  // Get unread email count, and print it on the LCD
  int tempCount = getEmailCount();
  lcd.print("You have ");
```

```
    lcd.print(tempCount);
    lcd.println(" unread mail.");

    if (tempCount != emailCount) // If it's a new value, update
    { // Do this to prevent blinking the same #
      emailCount = tempCount; // update emailCount variable
      printEmailCount(emailCount); // print the unread count
    }
  }

  // Bit of protection in case millis overruns:
  if (millis() < emailLastMillis)
    emailLastMillis = 0;
}
```

This is the function to get the e-mail count by executing the Python script:

```
int getEmailCount()
{
  int digits = 0;
  int temp[10];
  int emailCnt = 0;

  // Send a system call to run our python script and route the
  // output of the script to a file.
  system("python /home/root/emails.py > /media/card/emails.txt");

  File emailsFile = SD.open("emails"); // open emails for reading

  // read from emails until we hit the end or a newline
  while ((emailsFile.peek() != '\n') && (emailsFile.available()))
    temp[digits++] = emailsFile.read() - 48; // Convert from ASCII to
a number

  emailsFile.close(); // close the emails file
  system("rm /media/realroot/emails"); // remove the emails file

  // Turn the inidividual digits into a single integer:
  for (int i=0; i<digits; i++)
    emailCnt += temp[i] * pow(10, digits - 1 - i);

  return emailCnt;
}
```

Here, it is important to note that the whole code is based on the `system()` command, which allows the running of software (like Python) via the Arduino language. This is a really powerful function that allows you to execute more complex scripts than can be done with the Arduino IDE only: for example using Python. The following are the details for the MicroSD card:

```
void initSDCard()
{
  SD.begin();
}
```

The details to initialize the Ethernet connection are given here:

```
uint8_t initEthernet()
{
  if (Ethernet.begin(mac) == 0)
    return 0;
  else
    return 1;
}
```

Let's test everything, shall we?

The first thing to do is to put the Python script in the Galileo board with SCP. To do this, go to a terminal and type the following code:

```
scp emails.py root@192.168.1.103:.
```

If you don't want to do this, you can alternatively put the script at the root of the MicroSD card. However, to do this, you still need to modify the Arduino code by changing the path of the Python script, and then upload it again. Once done, you should see the number of unread e-mail counts. Here is how my entire project looked:

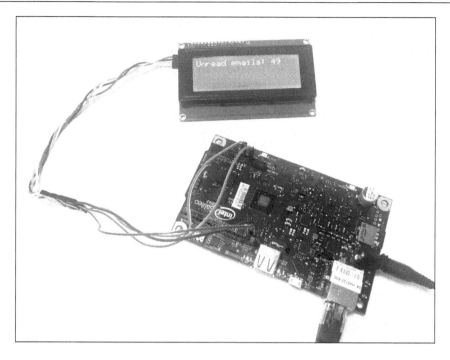

Here's a closer look at the LCD screen:

Summary

In this chapter, we looked at how to connect another existing application to our Galileo board. We connected the LCD screen to Arduino and created a code to integrate the e-mail data into the screen.

For future projects, you may opt to read other e-mail data from Gmail or connect to other web services. You can also integrate other apps into the board- for example, you can print new Tweets on the LCD (after all, a tweet is just 140 characters long, at most).

The next chapter is perfect for those with green thumbs. We will be automating our garden using the Galileo board.

9
Automated Remote Gardening with Intel Galileo

In this chapter, we will integrate an automation twist into gardening. Using a soil humidity sensor and a relay with a pump, we will set the water pump to water the plant whenever the humidity of the soil or the garden gets low.

Here is a brief summary of what we will do in this chapter:

- We will use a simple soil humidity sensor to measure the humidity of the soil or your garden
- We will set a trigger for the pump to water the plant if the humidity dips to a certain level
- We will connect everything to the Cloud to be monitored remotely

Hardware and software requirements

This project doesn't call for a new hardware requirement, as we will still use the same hardware that we have used in the previous chapters, where we measured data using the photocell and temperature sensor. We will just add a soil temperature and humidity sensor for this chapter.

The soil sensor that we will use is an SHT-10-based sensor encased in a metallic mesh. The casing keeps water from seeping into the insides where the sensor actually is, but allows air to pass through. This way, the sensor can measure the humidity and is kept dry.

This sensor contains a temperature/humidity sensor module from **Sensirion**, and the temperature readings have 0.5 percent precision, while the humidity readings have 4.5 percent precision.

 For more details about this gardening project enhancer, you may visit the following link:

https://www.adafruit.com/product/1298

Here is the image of the Soil Temperature/Moisture Sensor—SHT10:

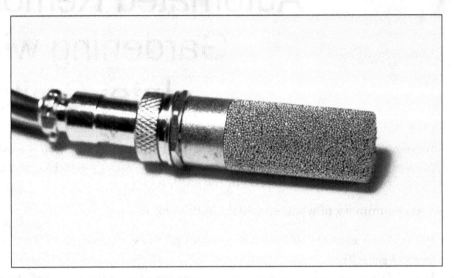

As for the relay, we will still use the 5V Pololu basic SPDT relay carrier, which we used in the previous projects.

For the pump, you can actually use a couple of commercially available carriers, which are already compatible with Arduino. We will now look at some options:

This 12V vacuum pump from SparkFun can be used for most small projects. It has a free flow range of 12 to 15 lpm. Head on to the following link for more details:

https://www.sparkfun.com/products/10398

Here is an image of a 12V vacuum pump:

Another pump that you can use is this 12V SEAFLO mini water pump. It uses a motor rated at 1.5A and can pump liquid at a rate of 350 gallons per hour; this too is from SparkFun:

https://www.sparkfun.com/products/10455

Here is an image of a 12V liquid pump:

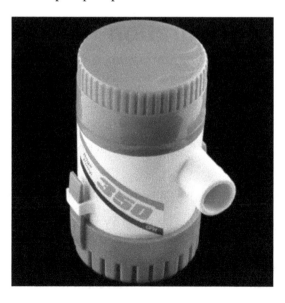

Another Arduino-ready pump is the DC liquid pump from TrossenRobotics. You will just need a RobotGeek large relay and a separate power supply to operate this. You can find out more about this pump, which can move up to 2 liters per minute, on the following link:

http://www.trossenrobotics.com/large-liquid-pump

Here is an image of a DC liquid pump:

Hardware configuration

As pointed out earlier, we can utilize the existing configuration used in the previous chapters for this project.

For the temperature and light sensors circuit, you can go back and refer to *Chapter 4, Monitoring Data Remotely.*

For relay integration, follow the set of relay instructions in *Chapter 3, Controlling Outputs Using the Galileo Board.*

When you completed these steps, we can add the pump and the soil sensor to them.

To add the soil sensor, connect the red wire to 5V pin on the board, green wire to ground, yellow wire to pin 7, and the blue wire to pin 6.

As for the pump, it really depends on which pump you are using. To connect the pump to the relay and then to the plant, you can consult your pump's data sheet.

Here is how my final hardware configuration looks:

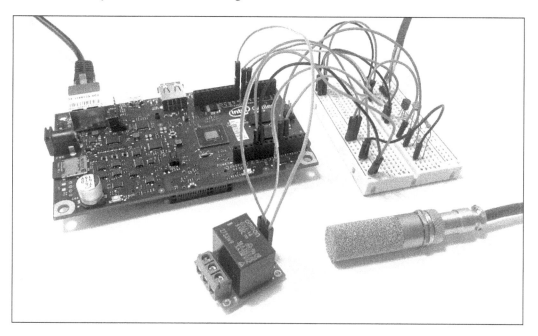

The next thing to do is to insert the soil humidity sensor into the soil of a potted plant or straight into your garden soil, depending on where you want to use it. I have planted the sensor in the soil of my potted plant.

Refer to the following image to see how I did it:

Testing all the sensors

At this point, we are ready to test things. We will start with all the sensors.

Again, we will use the Arduino IDE. Note that we will use the Sensirion library, which is required here to use the soil temperature and humidity sensor. Here is the complete code for this part:

```
// Libraries
#include <Sensirion.h>

// Pins
int temperature_pin = A1;
int light_pin = A0;

int dataPin = 6;
int clockPin = 7;

// Soil sensor instance
Sensirion soil_sensor = Sensirion(dataPin, clockPin);

void setup() {

  // Serial
  Serial.begin(115200);
}

// the loop routine runs over and over again forever:
void loop() {

  // Test photocell
  float light_reading = analogRead(light_pin);
  float light_level = light_reading/1024*100;
  Serial.print("Light level: ");
  Serial.println(light_level);

  // Test temperature sensor
  float temperature_reading = analogRead(temperature_pin);
  float temperature = (temperature_reading/1024*5000 - 500)/10;
  Serial.print("Temperature: ");
  Serial.println(temperature);

  // Test soil sensor
```

```
    uint16_t rawData;

    soil_sensor.measTemp(&rawData);
    float tempC = soil_sensor.calcTemp(rawData);
    soil_sensor.measHumi(&rawData);
    float humidity = soil_sensor.calcHumi(rawData, tempC);

    Serial.print("Soil humidity: ");
    Serial.println(humidity);

    Serial.println("");
    delay(1000);

}
```

The first step is to include the Sensirion library:

```
#include <Sensirion.h>
```

Then, we define the sensor's pins. You can name your sensor pins- I designated the temperature pin as A1 and the light pin as A0:

```
int temperature_pin = A1;
int light_pin = A0;

int dataPin = 6;
int clockPin = 7;
```

Next, we need to define the soil sensor instance:

```
Sensirion soil_sensor = Sensirion(dataPin, clockPin);
```

In the setup() function, we start the serial communication. This will be used for debugging purposes:

```
Serial.begin(115200);
```

Then, in the loop() function, we measure the light level, convert it to its digital equivalent, and print it:

```
float light_reading = analogRead(light_pin);
float light_level = light_reading/1024*100;
Serial.print("Light level: ");
Serial.println(light_level);
```

We then do the same for temperature. We get its analog value and convert it to Celsius:

```
float temperature_reading = analogRead(temperature_pin);
float temperature = (temperature_reading/1024*5000 - 500)/10;
Serial.print("Temperature: ");
Serial.println(temperature);
```

Next we measure the soil's humidity:

```
uint16_t rawData;

soil_sensor.measTemp(&rawData);
float tempC = soil_sensor.calcTemp(rawData);
soil_sensor.measHumi(&rawData);
float humidity = soil_sensor.calcHumi(rawData, tempC);

Serial.print("Soil humidity: ");
Serial.println(humidity);
```

We repeat the measurement process and print all three values parameters every second:

```
Serial.println("");
delay(1000);
```

Now, just upload the data to the board. After this, open the Arduino IDE serial monitor.

You will be able to see the results printing every second:

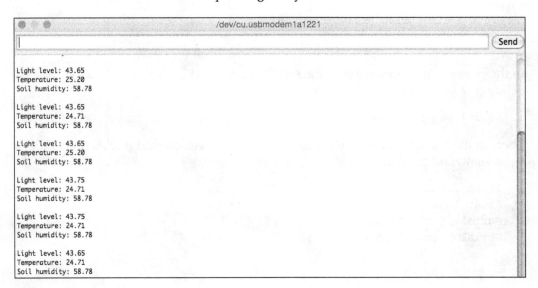

Automating your garden

After we have programmed the sensor controls, we will proceed to configure our water pump. In this next section, we will add the pump in the loop to automate gardening. Here is the complete code for this part:

```
// Libraries
#include <Sensirion.h>

// Pins
int temperature_pin = A1;
int light_pin = A0;

int dataPin = 6;
int clockPin = 7;

int relay_pin = 8;

// Soil sensor instance
Sensirion soil_sensor = Sensirion(dataPin, clockPin);

// Humidity thresholds
float low_threshold = 40;
float high_threshold = 60;

void setup() {

  // Serial
  Serial.begin(115200);

  // Pin mode
  pinMode(relay_pin, 8);
}

// the loop routine runs over and over again forever:
void loop() {

  // Measure & print data
  float light_reading = analogRead(light_pin);
  float light_level = light_reading/1024*100;
  Serial.print("Light level: ");
  Serial.println(light_level);

  float temperature_reading = analogRead(temperature_pin);
```

```
  float temperature = (temperature_reading/1024*5000 - 500)/10;
  Serial.print("Temperature: ");
  Serial.println(temperature);

  uint16_t rawData;

  soil_sensor.measTemp(&rawData);
  float tempC = soil_sensor.calcTemp(rawData);
  soil_sensor.measHumi(&rawData);
  float humidity = soil_sensor.calcHumi(rawData, tempC);

  Serial.print("Soil humidity: ");
  Serial.println(humidity);

  // Activate pump if humidity gets low
  if (humidity < low_threshold) {
    digitalWrite(relay_pin, HIGH);
  }

  // Deactivate pump if humidity gets high again
  if (humidity > high_threshold) {
    digitalWrite(relay_pin, LOW);
  }

  Serial.println("");
  delay(1000);

}
```

The first thing that we should do is define the relay pin:

```
int relay_pin = 8;
```

Next, define two humidity threshold levels. I used 40 percent for the low level and 60 percent for the high level. You may change the values if you want to:

```
float low_threshold = 40;
float high_threshold = 60;
```

We will also modify the relay pin as an output:

```
pinMode(relay_pin, 8);
```

We have now arrived at the core of this project.

What we want to do is to activate the pump if the humidity level gets low: that is when it reaches 40 percent in my case. The pump should stop once the humidity hits the upper threshold level, which is 60 in my example:

```
// Activate pump if humidity gets low
if (humidity < low_threshold) {
  digitalWrite(relay_pin, HIGH);
}

// Deactivate pump if humidity gets high again
if (humidity > high_threshold) {
  digitalWrite(relay_pin, LOW);
}
```

Upload the code to the board now.

You should be able to see the relay being activated and the pump watering your plant when the humidity gets too low. As mentioned earlier, you can adjust the threshold levels in your own set up as you find fit.

Congratulations, you have created your own automatic plant watering machine!

If things didn't work out, you can always check on a couple of things. First, check the hardware connections. You can also go back to *Chapter 3, Controlling Outputs Using the Galileo Board*, and *Chapter 4, Monitoring Data Remotely*, to verify that you made the right hardware connections.

Next, ensure that you have downloaded and uploaded the correct code to the board.

Connecting the project to the cloud

Now, we want to connect the data to the cloud so that we can monitor the project remotely. Once we are able to do this, we can check whether our plants need watering and initiate the watering process, even if we're not home.

To do this, we will first connect to Dweet.io to store our data.

We already saw how to use Dweet.io in *Chapter 6, Internet of Things with Intel Galileo*. Dweet.io is, as they claimed, the "Twitter for social machines". It publishes and subscribes data to your computer or any electronic device that connects to it, and it's fast and easy to use. No need for sign ups!

For more information about Dweet.io, you can visit the following link:
`http://dweet.io/`

What we need to do now is code the Arduino sketch. Here is the complete code for this part:

```
// Libraries
#include <Sensirion.h>
#include <SPI.h>
#include <Ethernet.h>

// MAC address
byte mac[] = {  0x98, 0x4F, 0xEE, 0x01, 0x90, 0xC5 };

// Ethernet client
EthernetClient client;
IPAddress server(54,173,147,83);

String relay_state = "Off";

// Dweet parameters
#define thing_name  "intel_galileo_53e5fx"

// Pins
int temperature_pin = A1;
int light_pin = A0;

int dataPin = 6;
int clockPin = 7;

int relay_pin = 8;

// Soil sensor instance
Sensirion soil_sensor = Sensirion(dataPin, clockPin);

// Humidity thresholds
float low_threshold = 40;
float high_threshold = 60;

void setup() {

  // Serial
```

```
  Serial.begin(115200);

  // Pin mode
  pinMode(relay_pin, 8);

  // start the Ethernet connection:
  if (Ethernet.begin(mac) == 0) {
    Serial.println("Failed to configure Ethernet using DHCP");
  }
}

// the loop routine runs over and over again forever:
void loop() {

  // Measure & print data
  float light_reading = analogRead(light_pin);
  float light_level = light_reading/1024*100;
  Serial.print("Light level: ");
  Serial.println(light_level);

  float temperature_reading = analogRead(temperature_pin);
  float temperature = (temperature_reading/1024*5000 - 500)/10;
  Serial.print("Temperature: ");
  Serial.println(temperature);

  uint16_t rawData;

  soil_sensor.measTemp(&rawData);
  float tempC = soil_sensor.calcTemp(rawData);
  soil_sensor.measHumi(&rawData);
  float humidity = soil_sensor.calcHumi(rawData, tempC);

  Serial.print("Soil humidity: ");
  Serial.println(humidity);

  // Activate pump if humidity gets low
  if (humidity < low_threshold) {
    relay_state = "On";
    digitalWrite(relay_pin, HIGH);
  }

  // Deactivate pump if humidity gets high again
  if (humidity > high_threshold) {
    relay_state = "Off";
```

```
      digitalWrite(relay_pin, LOW);
}


Serial.println("Connecting...");

// if you get a connection, report back via serial:
if (client.connect(server, 80)) {
  Serial.println("connected");

  // Make a HTTP request:
  Serial.print("Sending request... ");

  client.print("GET /dweet/for/");
  client.print(thing_name);
  client.print("?temperature=");
  client.print(temperature);
  client.print("&humidity=");
  client.print(humidity);
  client.print("&light=");
  client.print(light_level);
  client.print("&relay=");
  client.print(relay_state);
  client.println(" HTTP/1.1");

  client.println("Host: dweet.io");
  client.println("Connection: close");
  client.println("");

  Serial.println("done.");
}
else {
  // if you didn't get a connection to the server:
  Serial.println("connection failed");
}

// if there are incoming bytes available
// from the server, read them and print them:
if (client.connected()) {
  while (client.available()) {
    char c = client.read();
    Serial.print(c);
  }
}

// if the server's disconnected, stop the client:
```

```
client.stop();
Serial.println("");
Serial.println("Disconnecting.");

Serial.println("");
delay(5000);

}
```

The first step, as usual, is to include the libraries that we will use: Sensirion, SPI, and the Ethernet library:

```
#include <Sensirion.h>
#include <SPI.h>
#include <Ethernet.h>
```

Then get the Mac address of your Galileo board (which is written on the board itself) and encode it inside a variable:

```
byte mac[] = {  0x98, 0x4F, 0xEE, 0x01, 0x90, 0xC5 };
```

Next, create an Ethernet client and define the Dweet.io address:

```
EthernetClient client;
char * server = "dweet.io";
```

Set the default state of the relay to off:

```
String relay_state = "Off";
```

Define your device's name. It should be unique:

```
#define thing_name   "intel_galileo_53e5fx"
```

Then, in the setup() function, we start the connection:

```
if (Ethernet.begin(mac) == 0) {
  Serial.println("Failed to configure Ethernet using DHCP");
}
```

In the loop() function, we send data to Dweet.io:

```
if (client.connect(server, 80)) {
   Serial.println("connected");

   // Make a HTTP request:
   Serial.print("Sending request... ");

   client.print("GET /dweet/for/");
```

```
    client.print(thing_name);
    client.print("?temperature=");
    client.print(temperature);
    client.print("&humidity=");
    client.print("56.7");
    client.print("&light=");
    client.print(light_level);
    client.print("&relay=");
    client.print(relay_state);
    client.println(" HTTP/1.1");

    client.println("Host: dweet.io");
    client.println("Connection: close");
    client.println("");

    Serial.println("done.");
  }
```

Next, we read the response answer:

```
if (client.connected()) {
  while (client.available()) {
    char c = client.read();
    Serial.print(c);
  }
}
```

After this, we will stop the connection and set it to repeat every 5 seconds:

```
client.stop();
Serial.println("");
Serial.println("Disconnecting.");

Serial.println("");
delay(5000);
```

You can download the complete code from the following Github link:

```
https://github.com/marcoschwartz/galileo-blueprints-book
```

After downloading, upload the code to the Galileo board.

Next, open the serial monitor. You will be able to see the answers from Dweet.io. You will be get something like this:

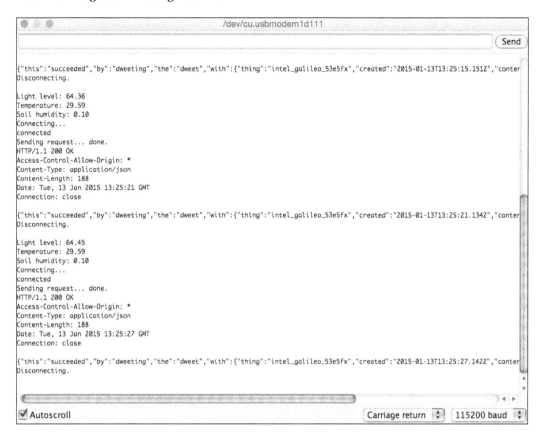

We can now store the data online and display it in a more comprehensible manner.

For this, we will use Freeboard, which is an online service that provides a dashboard for your Internet of Things data. We already used this in *Chapter 6, Internet of Things with Intel Galileo*. If you feed your data to Freeboard, it will create tables, graphs, and other appropriate displays of your data. It's simple to use and free.

To get going, go to the Freeboard link:

```
https://freeboard.io/
```

Create a freeboard account if you still don't have one, and if you do, just log in.

Next, add your Dweet.io data source and link it to your "thing", which was created in Dweet.io.

After this, add a panel with a Gauge object for the temperature readings. Link it to your temperature data source.

Repeat this for the light level and the soil humidity. Use the text widget for the relay state to simply show whether it's on or off.

You should be able to see the gauges point to their respective measurements. Here is how my dashboard looks:

Good job! You were able to build an automatic garden pump controlled by environmental factors, and you were able to pass the data to the cloud.

Summary

In this chapter, we built a completely automated system for plant gardening applications, based on the Intel Galileo. Here's a summary of what we did in this chapter:

First, we connected the soil humidity sensor and plugged it to the soil. Then, we used the relay to automate the watering of the plant. Finally, we connected the project to the cloud through Dweet.io for monitoring online. We also displayed the data on a dashboard through the Freeboard online service

In the next chapter, we will tackle a more advanced project. That is, we will build a complete home automation system! We will take everything you learned so far to build this system for your home using the Intel Galileo as a hub of your system.

10
Building a Complete Home Automation System

In the previous chapters, we used the Galileo board in standalone projects only, and sometimes, we connected it to web services to store and display data.

This time around, we will raise the bar to the next level. We will configure our Galileo board to function as a home automation hub, and we will have several devices connected to the Galileo board. We will be able to monitor them all from the Galileo board.

To describe this project further in a home automation context, we will display sensors in different parts of the house: for example, we will place a sensor in one room and a light output in another room.

To do this, we will first connect two modules based on Arduino. These modules need to be able to communicate via Wi-Fi so that we can access them remotely.

The first module will control a small LED and the other one will act like a sensor.

With the principles and techniques that you will learn in this project, you can extend the capabilities of your Galileo board and control your entire home.

Let's start!

Hardware and software requirements

First, let's see what hardware and software this project calls for.

As discussed, our Galileo boards need to be able to connect to our network. For this, we will need the Intel IoT image that we installed in *Chapter 4, Monitoring Data Remotely*. You may revisit that chapter to refresh your memory if you wish.

For our Arduino LED module, we will need the following major components:

1. The first one is the Arduino Uno board. This is the most used Arduino board at the present time, and it is the standard board for many projects.

2. The second board that we'll need is the Adafruit CC3000 Wi-Fi breakout board. This board utilizes **Serial Peripheral Interface (SPI)** for communication, so you can expect an easy data transfer and easy control of its data transfer speed. Asynchronous connections are made possible through this board's IRQ pin, providing you with a proper interrupt system.

3. The third major component is the Red LED with a 220 Ohm resistor.

These are the three major components that we will need for our Arduino board circuit. Other things that we will need are the breadboard and the jumper wires to create the connections.

You can read more about the boards and the other components that we will need through the following links:

- Arduino Uno board (`https://www.adafruit.com/products/50`)
- Red LED + 220 Ohm resistor
- Adafruit CC3000 Wi-Fi breakout board (`https://www.adafruit.com/product/1469`)
- Breadboard (`https://www.adafruit.com/products/64`)
- Jumper wires (`https://www.adafruit.com/products/1957`)

As for the sensor board, we will need another Arduino board, an Adafruit breakout board, a breadboard, and some jumper wires.

The additional component that we will utilize is the DHT11 sensor with a 10k Ohm resistor. The DHT11 sensor is a basic temperature and humidity digital sensor. It uses a capacitive sensor to detect humidity, and a thermistor to pick up the temperature of the surrounding air. This sensor is easy to use, but you need to be able to configure the timing properly to obtain the data.

For the communication side of things, it's better to have some background knowledge about aREST, since we will use this solution a lot.

aREST is an open source framework built for powerful applications based on Arduino and Raspberry Pi. It processes communications through serial, Wi-Fi, Ethernet, and many more. It can be easily used with Arduino and Pi, since it has the capability to include libraries for these boards. To install aREST, you can download it from the following link:

```
https://github.com/marcoschwartz/aREST
```

Then, simply place the folder that you downloaded inside your `Arduino /libraries` folder.

And to learn more about aREST, you can visit the following link:

```
http://arest.io/
```

Hardware configuration

We will now build the different boards based on the Arduino Uno, which we will later connect to the Galileo board via Wi-Fi.

Let's start the Arduino module with the LED light:

1. First, we need to connect the IRQ pin of the CC3000 board (that's the rightmost pin when you're facing the board) to pin 3 of the Arduino Uno.

2. Next, we will connect the VBAT pin to pin 5, and the CS pin to pin 10.

3. Then, we proceed to connecting the Adafruit SPI pins to the Arduino pins: MOSI, MISO, and CLK pins go to pins 11, 12, and 13, respectively.

4. Finally, we'll take care of the power supply. Adafruit's Vin should be connected to Arduino's 5V pin, and GND to GND.

5. The last things that we need to connect are the LED and resistor. Connect the resistor to pin 6 of the Uno.

6. Then, place the LED in series with the resistor. The negative pin of the LED should connect to the GND node.

You may refer to the following illustration for the entire configuration of the Arduino LED module:

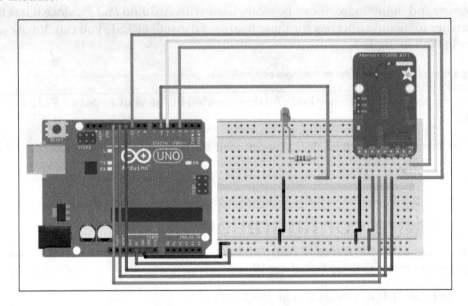

Here is how my module turned out after I connected everything:

Let's get working on the second module, which involves the DHT sensor. The connections are similar to the previous module, but we'll go over them again:

1. Again, we have to start with the IRQ pin. Connect this pin to pin 3 of the Arduino, the VBAT to pin 5, and CS pin to pin 10.

2. Next, we'll work on the SPI connections of the Adafruit board. Connect the MOSI, MISO, and CLK pins of the Adafruit to pins 11, 12, and 13 of the Arduino board, respectively.

3. Finally, we will configure the power supply. Connect the Vin pin to Arduino's 5V pin.

4. Connect the GND pins of both boards to each other.

5. As for the DHT sensor, here are its pin connections to the Arduino board: pin 1 to 5V, pin 4 to GND, and pin 2 to Arduino's pin 7.

6. Place the 10k Ohm resistor between pins 1 and 2.

You can see the entire configuration on the circuit illustration here:

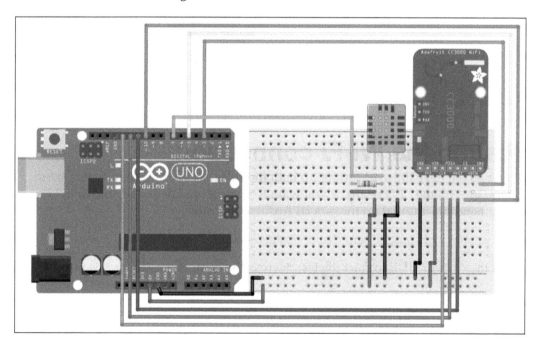

Check out how my module with the DHT sensor turned out:

Testing the Wi-Fi connection

Now that we have finished setting up both modules, we'll proceed to testing them. Let's start with the LED board. If you want to learn more about Arduino first, I recommend this tutorial on the official Arduino website:

`http://www.arduino.cc/en/Tutorial/HomePage`

Open the Arduino IDE to start the test. Here is the complete code for this part:

```
// Variables & functions
#define NUMBER_VARIABLES 2
#define NUMBER_FUNCTIONS 0

// Import required libraries
#include <Adafruit_CC3000.h>
#include <SPI.h>
#include <CC3000_MDNS.h>
#include <aREST.h>
```

```
#include <avr/wdt.h>

// DHT 11 sensor
#define DHTPIN 7
#define DHTTYPE DHT11

// These are the pins for the CC3000 chip if you are using a breakout
board
#define ADAFRUIT_CC3000_IRQ   3
#define ADAFRUIT_CC3000_VBAT  5
#define ADAFRUIT_CC3000_CS    10

// Create CC3000 instance
Adafruit_CC3000 cc3000 = Adafruit_CC3000(ADAFRUIT_CC3000_CS, ADAFRUIT_
CC3000_IRQ, ADAFRUIT_CC3000_VBAT);

// Create aREST instance
aREST rest = aREST();

// Your WiFi SSID and password
#define WLAN_SSID        "your_SSID"
#define WLAN_PASS        "your_password"
#define WLAN_SECURITY    WLAN_SEC_WPA2

// The port to listen for incoming TCP connections
#define LISTEN_PORT           80

// Server instance
Adafruit_CC3000_Server restServer(LISTEN_PORT);

void setup(void)
{
  // Start Serial
  Serial.begin(9600);

  // Give name and ID to device
  rest.set_id("2");
  rest.set_name("led");

  // Set up CC3000 and get connected to the wireless network.
  Serial.println("CC3000 init ...");
  if (!cc3000.begin())
  {
    while(1);
```

```
  }

  Serial.println("CC3000 init OK");

  if (!cc3000.connectToAP(WLAN_SSID, WLAN_PASS, WLAN_SECURITY)) {
    while(1);
  }
  while (!cc3000.checkDHCP())
  {
    delay(100);
  }

  // Start server
  restServer.begin();
  Serial.println(F("Listening for connections..."));

  displayConnectionDetails();

  wdt_enable(WDTO_4S);

}

void loop() {

  // Handle REST calls
  Adafruit_CC3000_ClientRef client = restServer.available();
  rest.handle(client);
  wdt_reset();

  // Check connection
  if(!cc3000.checkConnected()){while(1){}}
  wdt_reset();

}

// Print connection details of the CC3000 chip
bool displayConnectionDetails(void)
{
  uint32_t ipAddress, netmask, gateway, dhcpserv, dnsserv;

  if(!cc3000.getIPAddress(&ipAddress, &netmask, &gateway, &dhcpserv,
&dnsserv))
  {
    Serial.println(F("Unable to retrieve the IP Address!\r\n"));
```

```
    return false;
  }
  else
  {
    Serial.print(F("\nIP Addr: ")); cc3000.printIPdotsRev(ipAddress);
    Serial.print(F("\nNetmask: ")); cc3000.printIPdotsRev(netmask);
    Serial.print(F("\nGateway: ")); cc3000.printIPdotsRev(gateway);
    Serial.print(F("\nDHCPsrv: ")); cc3000.printIPdotsRev(dhcpserv);
    Serial.print(F("\nDNSserv: ")); cc3000.printIPdotsRev(dnsserv);
    Serial.println();
    return true;
  }
}
```

Let's examine the code step by step, so that you can fully understand it:

1. First, we will include the libraries that we need. These are the Adafruit CC3000, SPI, CC3000 MDNS, aREST, and the avr/wdt libraries.

   ```
   #include <Adafruit_CC3000.h>
   #include <SPI.h>
   #include <CC3000_MDNS.h>
   #include <aREST.h>
   #include <avr/wdt.h>
   ```

2. Next, we will define the IRQ, VBAT, and CS pins of the cc3000 chip:

   ```
   #define ADAFRUIT_CC3000_IRQ    3
   #define ADAFRUIT_CC3000_VBAT   5
   #define ADAFRUIT_CC3000_CS     10
   ```

3. Then, set your Wi-Fi network's name and password. You are free to replace the ID and password with your own:

   ```
   #define WLAN_SSID       "your_SSID"
   #define WLAN_PASS       "your_password"
   #define WLAN_SECURITY   WLAN_SEC_WPA2
   ```

4. We will also set the CC3000 instance:

   ```
   Adafruit_CC3000 cc3000 = Adafruit_CC3000(ADAFRUIT_CC3000_CS,
   ADAFRUIT_CC3000_IRQ, ADAFRUIT_CC3000_VBAT);
   ```

5. We will do the same for the aREST library:

   ```
   aREST rest = aREST();
   ```

6. Then define the port of the board:

   ```
   #define LISTEN_PORT        80
   ```

7. Next create a web server:

```
Adafruit_CC3000_Server restServer(LISTEN_PORT);
```

8. In the `setup()` function, we start the serial communication:

```
Serial.begin(9600);
```

9. Then, we give our preferred names to the boards. Its is best to give different names to your boards to avoid confusion. You can name one as `led` and the other one as `sensor`. It doesn't matter what name you give as long as you know which is which:

```
rest.set_id("2");
rest.set_name("led");
```

10. Then, we will connect it to the network:

```
Serial.println("CC3000 init ...");
  if (!cc3000.begin())
  {
    while(1);
  }

  Serial.println("CC3000 init OK");

  if (!cc3000.connectToAP(WLAN_SSID, WLAN_PASS, WLAN_SECURITY)) {
    while(1);
  }
  while (!cc3000.checkDHCP())
  {
    delay(100);
  }

  // Start server
  restServer.begin();
  Serial.println(F("Listening for connections..."));

  displayConnectionDetails();
```

11. Finally, in the `loop()` function, we will listen to requests:

```
Adafruit_CC3000_ClientRef client = restServer.available();
rest.handle(client);
```

Now that you have understood what each line of the code does, you can upload the code to your board. Once the upload is complete, open the serial monitor.

You will be able to see the IP address along with the other network details:

```
CC3000 init …
CC3000 init OK
Listening for connections …

IP Addr: 192.168.1.105
Netmask: 255.255.255.0
Gateway: 192.168.1.1
```

> Write down the IP address for later.

Do the same for the other module, which is the sensor board. You will find the code on the GitHub repository of the project.

Building the home automation interface

Now, we are ready to build the app that will facilitate the communication between the Galileo and Arduino boards.

We will use the same technique that we used in the previous chapter. We make use of the Node.js server to build a scalable network application, the Jade interface to manage interactions, and finally, JavaScript to link both.

Here is the first main.js file:

```
// Module
var express = require('express');
var app = express();

// Define port
var port = 3000;

// View engine
app.set('view engine', 'jade');
app.set('views', __dirname + '/views');

// Set public folder
app.use(express.static(__dirname + '/public'));

// Serve interface
```

```
app.get('/', function(req, res){
  res.render('interface');
});

// Rest
var rest = require("arest")(app);

// Add devices
rest.addDevice('http','192.168.1.105');
rest.addDevice('http','192.168.1.106');

// Start server
app.listen(port);
console.log("Listening on port " + port);
```

We will use the Express framework for this:

```
var express = require('express');
var app = express();
```

Then define the port of the app:

```
var port = 3000;
```

And identify `Jade` as the view engine:

```
// View engine
app.set('view engine', 'jade');
app.set('views', __dirname + '/views');

// Set public folder
app.use(express.static(__dirname + '/public'));
```

Next serve the interface file:

```
app.get('/', function(req, res){
  res.render('interface');
});
```

We use the Node.js and aREST modules to communicate with the boards.

Next, we will import the module:

```
var rest = require("arest")(app);
```

Then, we add the two Arduino boards. Ensure that you change the IP addresses to match your boards' IP addresses:

```
rest.addDevice('http','192.168.1.105');
rest.addDevice('http','192.168.1.106');
```

Finally, we start the server:

```
app.listen(port);
console.log("Listening on port " + port);
```

Change the package.json file with the aREST, express, and jade modules:

```
{
  "name": "Interface",
  "description": "",
  "version": "0.0.0",
  "main": "main.js",
  "engines": {
    "node": ">=0.10.0"
  },
  "dependencies": {
      "arest": "latest",
      "express": "latest",
      "jade": "latest"
  }
}
```

Now, we will code the interface to make it visually appealing. You can, of course, modify this code to tweak the project according to your project requirements.

Inside the Jade template file, here is how we'll configure the headers with a css file and JavaScript:

```
html
  head
    title Weather Station Interface
    script(src='https://code.jquery.com/jquery-2.1.1.min.js')
    script(src='/js/interface.js')
    link(rel='stylesheet', href="https://maxcdn.bootstrapcdn.com/
bootstrap/3.3.0/css/bootstrap.min.css")
```

We will then create indicators for the sensor board and add buttons to control the LED. You'll find the body of the page along with the temperature and humidity divisions, as shown here:

```
body
  .container
    h1 Home Automation Interface
    h3.row
      .col-md-4
        div Sensor reading
      .col-md-4
        div#temperature Temperature:
      .col-md-4
        div#humidity Humidity:
    h3.row
      .col-md-4
        div LED control
      .col-md-4
        button.btn.btn-lg.btn-block.btn-primary#on On
      .col-md-4
        button.btn.btn-lg.btn-block.btn-danger#off Off
```

We are now going to see the JavaScript code that will link this interface to our Node. js server. Here is the complete code for this part:

```
$( document ).ready(function() {

    // Refresh temperature & light level
    $.get('/sensor/temperature', function(json_data) {
      $('#temperature').html('Temperature: ' + (json_data.
temperature).toFixed(2) + ' C');
      $.get('/sensor/humidity', function(json_data) {
        $('#humidity').html('Humidity level: ' + (json_data.
humidity).toFixed(2) + ' %');
      });
    });

    // Relay control
    $.get('/led/mode/6/o');
    $('#on').click(function() {
      $.get('/led/digital/6/1');
    });

    $('#off').click(function() {
      $.get('/led/digital/6/0');
```

```
    });

  });
```

Meanwhile, we will update the sensor indicators in the JavaScript file by querying the sensor board:

```
$.get('/sensor/temperature', function(json_data) {
      $('#temperature').html('Temperature: ' + (json_data.
temperature).toFixed(2) + ' C');
        $.get('/sensor/humidity', function(json_data) {
          $('#humidity').html('Humidity level: ' + (json_data.
humidity).toFixed(2) + ' %');
        });
      });
```

After this, we will link the buttons to the right commands on the boards:

```
$.get('/led/mode/6/o');
    $('#on').click(function() {
      $.get('/led/digital/6/1');
    });

    $('#off').click(function() {
      $.get('/led/digital/6/0');
    });
```

You can now upload the code to the board and build it.

The server will start after a while. Once the server starts, you will be able to see both boards being discovered. You'll see something like this on your screen:

```
Intel XDK - Message Received: stop
=> Stopping App <=

UPLOADING: Uploading project bundle to IoT device.
[ Upload Complete ]
Intel XDK - Message Received: run
Listening on port 3000
Sending request to address: 192.168.1.105 for variable id
Sending request to address: 192.168.1.106 for variable id
Device added with ID: 2
Device added with ID: 1
```

Finally, go to your created interface by typing your Galileo board's IP address after `http://`. Here's an example:

`http://192.168.1.103:3000`

Ensure that you use the IP address of your own Galileo board. If not, the interface won't load correctly.

Once it loads, you will be able to see your home automation interface. My interface looked like this:

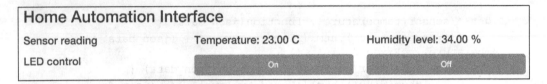

If you followed all the steps in this chapter, you'll see yours looking like this one too.

Now comes the cool part- try controlling the LEDs by clicking on the **On** and **Off** buttons.

The LED should light up and go off again as you control it:

Good job! You have now applied the basics in building your home automation system!

You can't just stop here though: the possibilities are endless when it comes to expanding your home system. You can add more indicators, control motors, and pumps. For instance, you can set a temperature range and turn on your heater once it falls below that range. You can also connect a digital clock and initiate your plant watering system during a certain time.

Take a look back at our past projects and explore ways to integrate them into your home automation hub.

In case the LED doesn't light up when you click on the button, you can go back and check on a few things, such as the hardware connections. You can also double check whether the software that you uploaded is the correct one. And of course, verify that you are using the correct IPs! Alternatively, you can test the LED with an external battery just to check whether the LED itself is not broken.

Summary

Let's summarize what we did in this chapter.

First, we built two Wi-Fi Arduino modules; one for the LED module and the other one for the sensor. Alternatively, you can also replace these Wi-Fi modules with other Galileo boards with Wi-Fi modules, or with Intel Edison boards that have built-in Wi-Fi.

Then, we tested them by pinging the boards and getting their IP addresses.

Finally, we integrated everything into a single interface with the help of the Node.js, a Jade interface, and lastly, JavaScript.

In the next chapter, we will use everything you learned so far for a different and more exciting application—building a mobile robot! You'll learn how to transform the Galileo board into a robot's brain.

11
Building a Mobile Robot Controlled by the Intel Galileo Board

In this chapter, we will use the Intel Galileo for a completely different application than we have so far. Indeed, we will use the board as be the "brain" of a mobile robot.

We are first going to see how to assemble the robot. We will mount the Galileo board on the robot, and also assemble an ultrasonic sensor on the robot.

Then, we will program the robot to do some basic tasks. First, we will do a complete test of the robot to check whether the Galileo board is correctly wired with the robot. Then, we will program the robot to perform a simple navigation task that is moving around while avoiding obstacles in front of the robot using the ultrasonic sensor. Let's dive in!

Hardware and software requirements

Let's first see what we need in this chapter to get started. Apart from the Galileo board, the most important piece of hardware here is the robot chassis. I used a simple DFRobot MiniQ 2 wheels chassis for this chapter. Here is an image of this chassis that comes with a base, two motors, and two wheels:

In the preceding image, you can see the ultrasonic sensor that was already mounted. Of course, you can use other robot chassis' for this chapter; they simply need to have two wheels with two motors, and you should be able to mount the Galileo board on it.

Then, we need an Arduino motor shield to connect the two DC motors. I used a DFRobot motor shield for this task. Note that I also used a prototyping for this project, but this is only necessary if you want to mount more sensors in the future.

We also need an ultrasonic sensor. I used an URM-37 sensor in this project, which is quite convenient to use:

Finally, you will need a battery to power the robot while it is moving around. I used a 7.2V battery from DFRobot, which can easily be put on the robot chassis under the Galileo board:

This battery doesn't come with a charger, but you can buy one from the following link:

`http://www.dfrobot.com/index.php?route=product/product&product_id=1212`

Here is a list of all the components that you will need for this chapter:

- DFRobot MiniQ 2 wheel chassis (`http://www.dfrobot.com/index.php?route=product/product&product_id=367`)
- DFRobot motor shield (`http://www.dfrobot.com/index.php?route=product/product&product_id=59`)
- URM37 ultrasonic sensor (`http://www.dfrobot.com/index.php?route=product/product&product_id=53`)
- 7.2V LiPo battery (`http://www.dfrobot.com/index.php?route=product/product&product_id=489`)

Hardware configuration

Let's now see how to assemble the robot:

1. First, I recommend that you mount the ultrasonic sensor directly on the robot chassis using an adapter, if possible. These usually come with the ultrasonic sensor.
2. Then, mount the Galileo board on top of the robot chassis using metal spacers, which usually come with the robot chassis.
3. After this, you can mount the motor shield on top of the Galileo board.
4. Also, connect the motors to the motor shield by ensuring that you connect both motors with the same polarity on the shield.

This is a view of the front of the robot at this stage:

At this stage, the back view of the robot is as follows:

5. Now, it's time to connect the ultrasonic sensor. Note that if you want to use a prototyping shield (so as to connect more sensors in the future), this is the time to do so. For the ultrasonic sensor, you have three pins to connect: VCC, GND, and the signal pin. When looking at the back of the sensor, this signal pin is the fourth pin starting from the left.

6. Connect the VCC pin to the shield 5V pin, GND to GND, and finally, the signal pin to the pin A0 of the shield. This is how your robot should look at this stage:

Note that, in this picture, a prototyping shield is also mounted because I want to extend the functionalities of the robot in the future.

Testing the robot

It's now time to test our robot. In this part, we will simply test all the functionalities of the robot with a simple sketch. We will make the robot go forward, stop, and turn left and right. We will also display the front distance from the sensor on the serial port.

Here is the complete code for this part:

```
// Motor pins
int speed_motor1 = 6;
int speed_motor2 = 5;
int direction_motor1 = 7;
int direction_motor2 = 4;

// Sensor pins
int distance_sensor = A0;

// Variable to be exposed to the API
int distance;

void setup(void)
{
  // Start Serial
  Serial.begin(115200);
}

void loop() {

  forward();
  delay(2000);
  left();
  delay(2000);
  right();
  delay(2000);
  stop();
  delay(2000);

  // Measure distance
  distance = measure_distance(distance_sensor);
  Serial.print("Measured distance: ");
  Serial.println(distance);

}

// Forward
int forward() {

  send_motor_command(speed_motor1,direction_motor1,100,1);
  send_motor_command(speed_motor2,direction_motor2,100,1);
  return 1;
```

```
}

// Backward
int backward() {

  send_motor_command(speed_motor1,direction_motor1,100,0);
  send_motor_command(speed_motor2,direction_motor2,100,0);
  return 1;
}

// Left
int left() {

  send_motor_command(speed_motor1,direction_motor1,75,0);
  send_motor_command(speed_motor2,direction_motor2,75,1);
  return 1;
}

// Right
int right() {

  send_motor_command(speed_motor1,direction_motor1,75,1);
  send_motor_command(speed_motor2,direction_motor2,75,0);
  return 1;
}

// Stop
int stop() {

  send_motor_command(speed_motor1,direction_motor1,0,1);
  send_motor_command(speed_motor2,direction_motor2,0,1);
  return 1;
}

// Function to command a given motor of the robot
void send_motor_command(int speed_pin, int direction_pin, int pwm,
boolean dir)
{
  analogWrite(speed_pin,pwm); // Set PWM control, 0 for stop, and 255
for maximum speed
  digitalWrite(direction_pin,dir);
}

// Measure distance from the ultrasonic sensor
```

```
int measure_distance(int pin){

   unsigned int Distance=0;
   unsigned long DistanceMeasured=pulseIn(pin,LOW);

   if(DistanceMeasured==50000){              // the reading is invalid.
       Serial.print("Invalid");
    }
     else{
        Distance=DistanceMeasured/50;     // every 50us low level
stands for 1cm
     }

   return Distance;
}
```

Let's now see the details of this code. It starts by declaring the pins to which the motor shield is connected to:

```
int speed_motor1 = 6;
int speed_motor2 = 5;
int direction_motor1 = 7;
int direction_motor2 = 4;
```

Then, we also declare a pin for the ultrasonic sensor, which is connected to the analog pin A0:

```
int distance_sensor = A0;
```

We also declare a variable that will store the distance in front of the sensor:

```
int distance;
```

Then, in the loop() function of the sketch, we make the robot go forward for a while. Then, we make it go left, right, and stop:

```
forward();
delay(2000);
left();
delay(2000);
right();
delay(2000);
stop();
delay(2000);
```

We will look at the details of these functions later in this chapter. Still in the `loop()` function, we then measure the distance in front of the robot and display it inside the serial monitor:

```
distance = measure_distance(distance_sensor);
Serial.print("Measured distance: ");
Serial.println(distance);
```

We will also look at the details of this function later in this chapter.

First, let's look at the details for one of the functions that makes the robot move around, for example, the `forward()` function:

```
int forward() {

    send_motor_command(speed_motor1,direction_motor1,100,1);
    send_motor_command(speed_motor2,direction_motor2,100,1);
    return 1;
}
```

You can see that this function calls another function that we actually call twice, once for each motor.

These are the details of the function that we used to control a given motor:

```
void send_motor_command(int speed_pin, int direction_pin, int pwm,
boolean dir)
{
    analogWrite(speed_pin,pwm); // Set PWM control, 0 for stop, and 255
for maximum speed
    digitalWrite(direction_pin,dir);
}
```

It basically consists of applying the right speed command via the `analogWrite()` function, and then setting the direction of the motor.

Finally, let's see the details of the function that we used to measure the distance in front of the robot:

```
int measure_distance(int pin){

    unsigned int Distance=0;
    unsigned long DistanceMeasured=pulseIn(pin,LOW);

    if(DistanceMeasured==50000){                  // the reading is invalid.
        Serial.print("Invalid");
    }
```

```
    else{
        Distance=DistanceMeasured/50;       // every 50us low level
    stands for 1cm
    }

    return Distance;
}
```

This function is based on the `pulseIn()` function of Arduino, which is used to automatically measure the width of the pulse returned by such sensors. As we know that the width of this pulse is proportional to the distance in front of the sensor, we can easily calculate the distance in front of the sensor.

As usual, you will find the code for this on the GitHub repository of this project.

It's now time to test this code. First, ensure that the robot is put on a platform where the wheels don't touch the ground, or you will run into problems quite quickly.

Then, open the code with the Arduino IDE, connect the board to your computer via a micro USB cable (ensuring that the main 12V power is connected to the board as well) and upload the code. After this, open the serial monitor.

You should immediately see that the robot wheels are turning, executing the different behaviors that we coded. You should also see that the distance in front of the robot is being printed on the serial monitor. If this doesn't work, there are several things you can check. First, ensure that you have connected all the cables between the motor shield and the motors correctly. The polarity of the motors can also be wrong, so ensure that you connected both motors with the same polarity. Also check that you are powering your Galileo board from the 12V power adapter, otherwise, it won't have enough power to move the wheels.

Autonomous navigation

Now that we are sure that our robot is correctly wired, it's time to write an application for moving it around while avoiding obstacles in front of the robot. Here is the complete code for this part:

```
// Motor pins
int speed_motor1 = 6;
int speed_motor2 = 5;
int direction_motor1 = 7;
int direction_motor2 = 4;

// Sensor pin
```

```
int distance_sensor = A0;

void setup(void)
{
  // Start Serial
  Serial.begin(115200);

  // Go forward by default
  forward();
}

void loop() {

  // Measure distance
  int distance = measure_distance(distance_sensor);

  // If obstacle in front
  if (distance < 10) {

    // Go backward
    backward();
    delay(2000);

    // Turn around
    left();
    delay(500);

    // Go forward again
    forward();
  }

}

// Forward
int forward() {

  send_motor_command(speed_motor1,direction_motor1,200,1);
  send_motor_command(speed_motor2,direction_motor2,200,1);
  return 1;
}

// Backward
```

```
int backward() {

  send_motor_command(speed_motor1,direction_motor1,200,0);
  send_motor_command(speed_motor2,direction_motor2,200,0);
  return 1
}


// Left
int left() {

  send_motor_command(speed_motor1,direction_motor1,150,0);
  send_motor_command(speed_motor2,direction_motor2,150,1);
  return 1;
}


// Right
int right() {

  send_motor_command(speed_motor1,direction_motor1,150,1);
  send_motor_command(speed_motor2,direction_motor2,150,0);
  return 1;
}


// Stop
int stop() {

  send_motor_command(speed_motor1,direction_motor1,0,1);
  send_motor_command(speed_motor2,direction_motor2,0,1);
  return 1;
}


// Function to command a given motor of the robot
void send_motor_command(int speed_pin, int direction_pin, int pwm,
boolean dir)
{
  analogWrite(speed_pin,pwm); // Set PWM control, 0 for stop, and 255
for maximum speed
  digitalWrite(direction_pin,dir);
}


// Measure distance from the ultrasonic sensor
int measure_distance(int pin){

  unsigned int Distance=0;
```

```
    unsigned long DistanceMeasured=pulseIn(pin,LOW);

    if(DistanceMeasured==50000){              // the reading is invalid.
        Serial.print("Invalid");
    }
     else{
        Distance=DistanceMeasured/50;         // every 50us low level
  stands for 1cm
    }

    return Distance;
}
```

We will now look the details of this code. You will notice that a lot of common parts can be seen that a lot of pieces of the code are similar to the previous part, so we will only go through the differences here.

In the `setup()` function of the sketch, we make the robot go forward by default:

```
    forward();
```

Then, in the `loop()` function, we continuously measure the distance in front of the sensor and store it in a variable:

```
    int distance = measure_distance(distance_sensor);
```

If the distance in front of the sensor is less than 10 cms, we make the robot go backward for 2 seconds, then turn left for half a second, and then go forward again. This is done by the following piece of code:

```
    if (distance < 10) {

        // Go backward
        backward();
        delay(2000);

        // Turn around
        left();
        delay(500);

        // Go forward again
        forward();
    }
}
```

As usual, you will find the code for this on the GitHub repository of this project.

It's now time to test the code for this part. Again, ensure that the robot is connected to the 12V power adapter, and that it is mounted on some kind of platform so that the wheels don't touch the ground.

Then, load the code into the Arduino IDE and upload the code to the robot. Then, you can disconnect the USB cable and the power adapter, and plug the robot battery into the main plug of the Galileo board. You can store the battery just below the Galileo board.

Once this is done, just place the robot on the ground, and let it move around. You will see that it is continuously moving forward until it encounters an obstacle. If this is the case, you will notice that it will execute the obstacle avoidance behavior that we defined earlier.

Summary

We used the Intel Galileo board to act as the "brain" of a mobile robot. After building the robot, we tested it with a simple program, to ensure that all the connections were alright. Then, we built a more complex behavior using the Galileo board by making the robot avoid obstacles while moving around.

There are, of course, many ways to improve this project. For example, you can connect more sensors to the robot. One way would be to connect a digital compass to the robot and use it to get the direction that the robot is facing toward. That way, you can program the robot to turn from a given angle, without having to guess the speed of its wheels.

In the next chapter, you will see how to use the MQTT protocol on your Galileo board, which will allow you to control your Galileo projects in real time from anywhere in the world.

12

Controlling the Galileo Board from the Web in Real Time Using MQTT

In the last chapter of this book, we will see how to control our Galileo board in real time from anywhere in the world. This project is more complex than the other projects in this book as it uses a new protocol, the **Message Queuing Telemetry Transport** (**MQTT**), which we will use to control our Galileo board.

However, we will use an online service called **PubNub** to make the process much easier. This will make it really easy to control our board remotely and to monitor data from the board. As an example, we will connect two sensors to the Galileo board: a motion sensor and a light level sensor. We will see how to monitor the state from these two sensors in real time from anywhere in the world.

We will also connect a relay to the Galileo board that can be used to control a lamp, for example. We will see how to control this relay in real time from anywhere using the MQTT protocol.

Hardware and software requirements

Let's first see what is needed for this project. You need, of course, the same Intel Galileo Gen 2 board that we used for all the projects in this book.

You will need a 5V relay as well: I used a Pololu 5V relay for this task, and you will also need a photocell along with a 10k Ohm resistor and a standard 5V PIR motion sensor. Finally, you will need the usual breadboard and jumper wires. Here is a list of all the components that you will need for this project:

- Intel Galileo Gen 2 board (`https://www.sparkfun.com/products/13096`)
- PIR motion sensor (`https://www.sparkfun.com/products/8630`)
- Photocell (`https://www.sparkfun.com/products/9088`)
- 10k Ohm resistor (`https://www.sparkfun.com/products/8374`)
- Pololu 5V relay module (`http://www.pololu.com/product/2480`)
- Breadboard (`https://www.sparkfun.com/products/12002`)
- Jumper wires (`https://www.sparkfun.com/products/8431`)

On the software side, we will assume that your Galileo board is ready to be used with the Intel IoT image installed on a MicroSD card. Please refer to any of the previous chapters, for example *Chapter 1, Setting Up the Galileo Board and the Development Environment*, if this is not the case.

Hardware configuration

We are now going to see how to configure the hardware for this chapter. Let's start by connecting the power supply; connect to the 5V pin on the board coming from the Galileo board to the red power line on the breadboard, and the GND pin of the Galileo board to the blue power line of the breadboard.

Then, we will connect the relay. Connect the VCC pin of the relay to the red power line, GND to the blue power line, and the SIG pin of the relay to pin number 7.

For the PIR motion sensor, the process is actually very similar. Connect the VCC pin to the red power line, GND to the blue power line, and the OUT pin of the sensor to pin number 8.

Finally, place the photocell in the series with the 10k Ohm resistor on the breadboard. Then, connect the free pin of the photocell to the red power line, and the free pin of the resistor to the blue power line. Also, connect the middle pin to the analog pin A0 of the Galileo board.

The following image shows the final result:

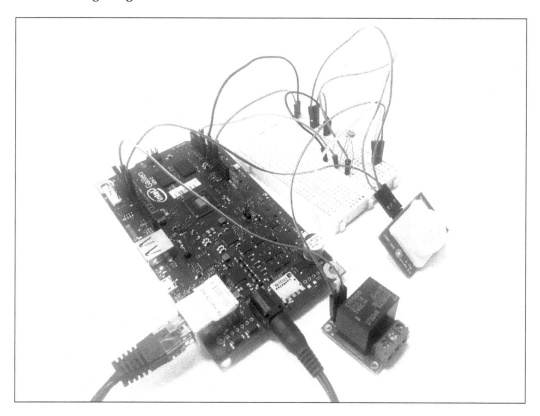

The MQTT protocol

Before going further, we will go into a bit more detail about the MQTT protocol that we are going to use in this project. You can find all the details about this protocol at the following link:

`http://mqtt.org/`

The MQTT protocol was created as a lightweight messaging protocol for the Internet of Things. It is now used in many IoT projects as well as **machine-to-machine (M2M)** applications.

It works on the publish/subscribe principle. An MQTT device can subscribe to a given channel; this device will then receive all the messages posted on this channel. This is what we are going to use to receive orders from the Web to control the relay.

On the other hand, an MQTT device can also publish messages to a given channel. This is what we are going to use to send sensor measurements and monitor them from the Web.

Creating a PubNub account

One way to use MQTT to control our Galileo board in real time would be to install our own MQTT server (called a broker) online, and control our board from there.

However, this is really complex. We would first have to build the server locally, then find a host for our server, and finally upload the server online.

Instead of all this, we will use PubNub, where you can simply create an account to control your devices remotely using their libraries, which are based on MQTT.

Here are the steps for creating an account in PubNub:

1. The first step is to create an account by visiting the following link:

   `http://www.pubnub.com/`

2. You will arrive on their main page where you can enter your contact details, as shown in the following screenshot:

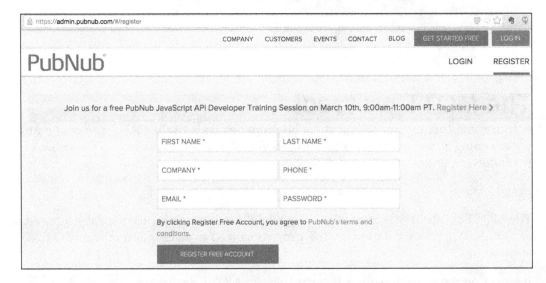

3. Then, you will be required to create a new application. I simply named mine **Galileo**:

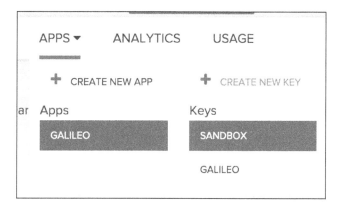

4. After this, you can also create a new key from the same menu. Click on your application to see the details:

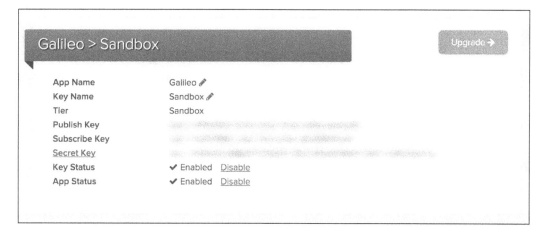

5. You can see that there are many keys here that I blurred out. You will need the **Publish key** and the **Subscribe key** for our application, as we will both send data (sensor's measurements) and receive data (the commands for our relays).

Write down these two keys somewhere, as we will need them soon.

Writing the code for the project

We will now look at the code for the project. Once again, we will use Node.js as the base of our application, and we will code it inside the Intel XDK software that we already used in previous chapters.

Here is the complete code for this part:

```
// Includes
var mraa = require("mraa");
var util = require('util');

// Sensors & relay pins
var light_sensor_pin = new mraa.Aio(0);

var motion_sensor_pin = new mraa.Gpio(8);
motion_sensor_pin.dir(mraa.DIR_IN);

var relay_pin = new mraa.Gpio(7);
relay_pin.dir(mraa.DIR_OUT);

// PubNub client
var pubnub = require("pubnub")({
    ssl             : true,   // <- enable TLS Tunneling over TCP
    publish_key     : "your_publish_key",
    subscribe_key   : "your_subscribe_key"
});

// Function to publish light level
function send_light_level() {

  // Measure light level
  var a = light_sensor_pin.read();
  var light_level = a/1024*100;
  light_level = light_level.toPrecision(4);

  // Send it to PubNub
  var message = { "level" : light_level};
  pubnub.publish({
      channel   : 'light',
      message   : message,
      callback  : function(e) { console.log( "SUCCESS!", e ); },
```

```
        error     : function(e) { console.log( "FAILED! RETRY PUBLISH!",
e ); }
    });

}

// Repeat every 10 seconds
send_light_level();
setInterval(send_light_level, 10000);

// Check motion sensor data every second
var motion_sensor_state = false;
setInterval(function() {

  // Check current state of motion sensor
  var current_state = motion_sensor_pin.read();

  // If different, publish a message
  if(current_state != motion_sensor_state) {
    motion_sensor_state = current_state;
    message = { "motion_sensor" : false};

    if (motion_sensor_state == 0) {
      message = { "motion_sensor" : false};
    }
    else {
      message = { "motion_sensor" : true};
    }

    pubnub.publish({
      channel   : 'motion',
      message   : message,
      callback  : function(e) { console.log( "SUCCESS!", e ); },
      error     : function(e) { console.log( "FAILED! RETRY PUBLISH!",
e ); }
    });
  }

}, 1000);

// Listen for messages to control the relay
pubnub.subscribe({
  channel  : "relay",
```

```
callback : function(message) {
  console.log( " > ", message );

  // If we have a relay field in the message, act accordingly
  if (message.relay) {
    if (message.relay == 'on') {
      relay_pin.write(1);
    }
    if (message.relay == 'off') {
      relay_pin.write(0);
    }
  }
}
});
```

Let's now look at the detail of the code. It starts by including the required libraries to control the pins of the Galileo board:

```
var mraa = require("mraa");
var util = require('util');
```

Then, we declare that we have an analog sensor on the pin A0:

```
var light_sensor_pin = new mraa.Aio(0);
```

After this, we declare the pin of the motion sensor and set this pin as an input pin using the `dir()` function:

```
var motion_sensor_pin = new mraa.Gpio(8);
motion_sensor_pin.dir(mraa.DIR_IN);
```

We do the same for the relay pin, setting this one as an output:

```
var relay_pin = new mraa.Gpio(7);
relay_pin.dir(mraa.DIR_OUT);
```

We can then create the instance of the PubNub library that we will use to send and receive messages from the PubNub service. This is also where you have to insert the keys that you got from the PubNub website:

```
var pubnub = require("pubnub")({
    ssl            : true,  // <- enable TLS Tunneling over TCP
    publish_key    : "your_publish_key",
    subscribe_key  : "your_subscribe_key"
});
```

Also, note that you need to modify the package.json file, which is placed in the same folder as the main Node.js file, in order to include the PubNub library so that it is automatically installed when you transfer the project to the Galileo board.

We can now create the function to send the current light level to the PubNub service. Here is the complete code for this function:

```
function send_light_level() {

    // Measure light level
    var a = light_sensor_pin.read();
    var light_level = a/1024*100;
    light_level = light_level.toPrecision(4);

    // Send it to PubNub
    var message = { "level" : light_level};
    pubnub.publish({
        channel    : 'light',
        message    : message,
        callback   : function(e) { console.log( "SUCCESS!", e ); },
        error      : function(e) { console.log( "FAILED! RETRY PUBLISH!",
e ); }
    });
}
```

This function starts by measuring the light level and converting it to a percentage, just as we did earlier in this book.

Then, we call the PubNub library, and send a message to the PubNub server via the publish command. Note that we also create a new channel on PubNub that will only contain light level measurements.

Now, we can execute this function and repeat it every 10 seconds:

```
send_light_level();
setInterval(send_light_level, 10000);
```

Let's now see how the motion sensor works. Unlike the light level, we only want to transmit data if a change is detected at the level of the motion sensor. It is indeed useless to send new data if no motion is detected. This is why we will use the JavaScript setInterval() function here to regularly check the state of the motion sensor.

Here is the complete code for this part:

```
var motion_sensor_state = false;
setInterval(function() {

    // Check current state of motion sensor
    var current_state = motion_sensor_pin.read();

    // If different, publish a message
    if(current_state != motion_sensor_state) {
      motion_sensor_state = current_state;
      message = { "motion_sensor" : false};

      if (motion_sensor_state == 0) {
        message = { "motion_sensor" : false};
      }
      else {
        message = { "motion_sensor" : true};
      }

      pubnub.publish({
         channel   : 'motion',
         message   : message,
         callback  : function(e) { console.log( "SUCCESS!", e ); },
         error     : function(e) { console.log( "FAILED! RETRY PUBLISH!",
   e ); }
      });
    }

}, 1000);
```

In this function, we can see that the state of the motion sensor is checked every second. Then, only if the state has changed compared to the previous reading, we send a message to PubNub using the publish function on the `motion` channel.

For the relay, we will use the `subscribe` function. This will ensure that we can receive messages on the `relay` channel, which we will use to control the relay remotely. Here is the complete code for this function:

```
pubnub.subscribe({
   channel   : "relay",
   callback  : function(message) {
```

```
    console.log( " > ", message );

    // If we have a relay field in the message, act accordingly
    if (message.relay) {
      if (message.relay == 'on') {
        relay_pin.write(1);
      }
      if (message.relay == 'off') {
        relay_pin.write(0);
      }
    }
  }
});
```

We can see here that we first need to subscribe to the channel. Then, if a message is received, we check whether it concerns our relay, and then activate or deactivate the relay accordingly.

 Note that you can find all the code for this project on the GitHub repository of this book:

`https://github.com/marcoschwartz/galileo-blueprints-book`

We can now test the project. For this, you need to perform the following steps:

1. First, upload the project to the board using the Intel XDK software that we used earlier.

2. Ensure that the files are uploaded to the Galileo board, and that all the Node. js modules are installed.

3. Then, to monitor the different MQTT channels that we use in this project, you can use the PubNub console, which is very convenient. You can access it at the following link:

   `http://www.pubnub.com/console/`

This is what you will get:

4. Note that here I already inserted my own subscribe and publish keys, which I blurred out in this image. You also have to insert your own subscribe and publish keys now.

5. Then, connect to the channel called **light** to check the light level measurements. You will see that every 10 seconds, a message will be published inside that channel containing the light level data:

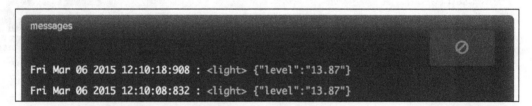

6. Now, still using the console, check the channel called **motion** to check the data coming from the motion sensor. Also, hover your hand in front of the motion sensor to activate it. Every time the motion sensor changes its state, you will see a message being sent to this channel:

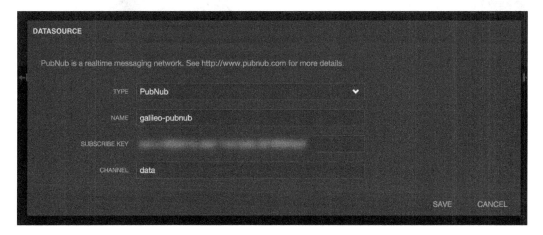

This means that our system is working correctly. However, this is really not convenient if you want to monitor your data. That is why, in the next part, we will look at how to integrate our project with an Internet of Things dashboard- the Freeboard.io, which we have already used in this book. That way, we will be able to visualize our data graphically.

Monitoring your board remotely

We will now see how to integrate our project based on PubNub with Freeboard.io, which we already used earlier in the project. I decided to use Freeboard.io again here because it integrates well with PubNub. Visit the following link for more information:

```
https://www.freeboard.io/
```

You will be prompted to create a new account. If you don't already have one, create a new dashboard, and a new **DATASOURCE**:

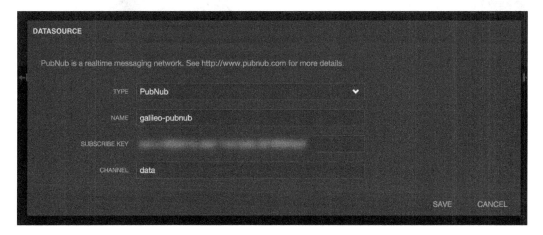

We only want to receive data, so you will only need your subscribe key here. Create two data sources, one for the "light" channel and another for the "motion" channel. This is the datasource for the **motion** channel:

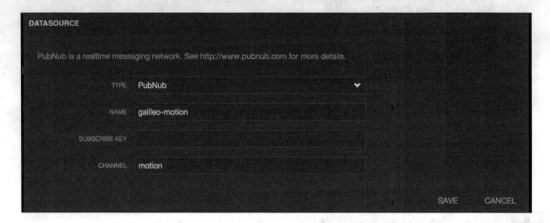

Now we need to create some widgets and link them to the given data sources in order to display the data.

For the light level, we will create a gauge indicator. Simply link it to the **light** datasource that you created earlier, and use a range going from **0** to **100**:

For the motion sensor, I simply used an **Indicator Light** widget and linked it to the **motion** data source that I created earlier:

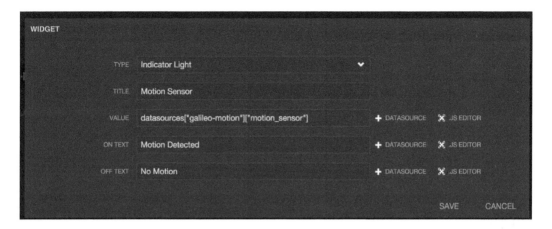

The following is the final result:

You can see that this dashboard is updated in real time as data comes from the Galileo board to the PubNub service.

Controlling the relay remotely

We still have to see how to control the relay remotely, so for this, let's go back to the PubNub console that we created earlier.

Now, connect the `relay` channel just as you did earlier with the other channels. We don't want to receive the data here, but we will send some data.

For this, we will use the message field, where you can type a message and send it to our board. Simply send the following message to turn the relay on:

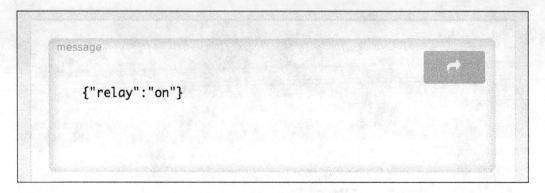

You can immediately see and hear that the relay is on. To turn it off again, type the following message:

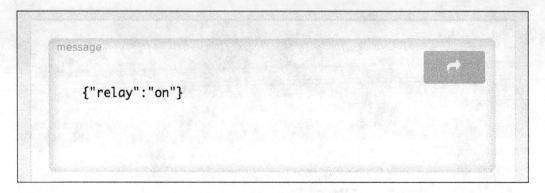

Congratulations, you can now control your relay from anywhere in the world!

Summary

First, we built the hardware for this project: basically two sensors and a relay that were connected to the Galileo board. Then, we saw what the MQTT protocol is, and how we can easily integrate it into our project using PubNub.

After that, we wrote the code for our project and tested it with the PubNub console. We also covered how to display all the measurements inside a web dashboard. Finally, we saw how to control the relay in real time from anywhere in the world.

There are many ways to improve on this project. You can, for example, integrate more sensors on your Galileo board and display their measurements in real time on the dashboard you created at Freeboard.io. You can also code your own dashboard, for example, integrating an on/off button to control the relay remotely.

We are finally at the end of this book. I hope that you enjoyed reading it as much as I enjoyed the writing process! It is now time for you to take what you learned in this book and start building your own projects. There are no limits to what you can make with the Galileo board, especially in the Internet of Things field. I also invite you to regularly check the Intel Galileo pages to stay informed of the latest news about the board, as the platform is always evolving and getting better. Have fun!

Thank you for buying
Intel Galileo Blueprints

About Packt Publishing

Packt, pronounced 'packed', published its first book, *Mastering phpMyAdmin for Effective MySQL Management*, in April 2004, and subsequently continued to specialize in publishing highly focused books on specific technologies and solutions.

Our books and publications share the experiences of your fellow IT professionals in adapting and customizing today's systems, applications, and frameworks. Our solution-based books give you the knowledge and power to customize the software and technologies you're using to get the job done. Packt books are more specific and less general than the IT books you have seen in the past. Our unique business model allows us to bring you more focused information, giving you more of what you need to know, and less of what you don't.

Packt is a modern yet unique publishing company that focuses on producing quality, cutting-edge books for communities of developers, administrators, and newbies alike. For more information, please visit our website at www.packtpub.com.

About Packt Open Source

In 2010, Packt launched two new brands, Packt Open Source and Packt Enterprise, in order to continue its focus on specialization. This book is part of the Packt Open Source brand, home to books published on software built around open source licenses, and offering information to anybody from advanced developers to budding web designers. The Open Source brand also runs Packt's Open Source Royalty Scheme, by which Packt gives a royalty to each open source project about whose software a book is sold.

Writing for Packt

We welcome all inquiries from people who are interested in authoring. Book proposals should be sent to author@packtpub.com. If your book idea is still at an early stage and you would like to discuss it first before writing a formal book proposal, then please contact us; one of our commissioning editors will get in touch with you.

We're not just looking for published authors; if you have strong technical skills but no writing experience, our experienced editors can help you develop a writing career, or simply get some additional reward for your expertise.

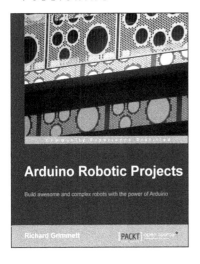

Arduino Robotic Projects

ISBN: 978-1-78398-982-9   Paperback: 240 pages

Build awesome and complex robots with the power of Arduino

1. Develop a series of exciting robots that can sail, go under water, and fly.

2. Simple, easy-to-understand instructions to program Arduino.

3. Effectively control the movements of all types of motors using Arduino.

4. Use sensors, GSP, and a magnetic compass to give your robot direction and make it lifelike.

Arduino Development Cookbook

ISBN: 978-1-78398-294-3   Paperback: 246 pages

Over 50 hands-on recipes to quickly build and understand Arduino projects, from the simplest to the most extraordinary

1. Get quick, clear guidance on all the principle aspects of integration with the Arduino.

2. Learn the tools and components needed to build engaging electronics with the Arduino.

3. Make the most of your board through practical tips and tricks.

Please check **www.PacktPub.com** for information on our titles

www.ingramcontent.com/pod-product-compliance
Lightning Source LLC
Chambersburg PA
CBHW060130060326
40690CB00018B/3826